Patrick Wüstefeld

Structure and Diagenesis in Upper Carboniferous Tight Gas Reservoirs in NW Germany

Structure and Diagenesis in Upper Carboniferous Tight Gas Reservoirs in NW Germany

by
Patrick Wüstefeld

KIT Scientific Publishing

Karlsruher Institut für Technologie
Institut für Angewandte Geowissenschaften

Structure and Diagenesis in Upper Carboniferous
Tight Gas Reservoirs in NW Germany

Zur Erlangung des akademischen Grades eines Doktors der Naturwissenschaften von der KIT-Fakultät für Bauingenieur-, Geo- und Umweltwissenschaften des Karlsruher Instituts für Technologie (KIT) genehmigte Dissertation

von Patrick Wüstefeld, M.Sc. RWTH aus Haan

Tag der mündlichen Prüfung: 21. Juli 2017
Gutachter: Prof. Dr. Christoph Hilgers, Prof. Dr. Janos Urai

Impressum

KIT Scientific Publishing

Karlsruher Institut für Technologie (KIT)
KIT Scientific Publishing
Straße am Forum 2
D-76131 Karlsruhe

KIT Scientific Publishing is a registered trademark
of Karlsruhe Institute of Technology.
Reprint using the book cover is not allowed.

www.ksp.kit.edu

Print on Demand 2018 – Gedruckt auf FSC-zertifiziertem Papier

ISBN 978-3-7315-0734-5
DOI 10.5445/KSP/1000076144

Structure and Diagenesis in Upper Carboniferous Tight Gas Reservoirs in NW Germany

zur Erlangung des akademischen Grades eines

Doktors der Naturwissenschaften

von der Fakultät für Bauingenieur-, Geo- und Umweltwissenschaften

des Karlsruher Instituts für Technologie (KIT)

genehmigte

Dissertation

von

Patrick Wüstefeld, M.Sc. RWTH

aus Haan

Tag der mündlichen Prüfung: 21. Juli 2017
Erster Gutachter: Prof. Dr. Christoph Hilgers
Zweiter Gutachter: Prof. Dr. Janos Urai

Acknowledgments

First of all, I am sincerely grateful to my supervisor, Prof. Dr. Christoph Hilgers (former RWTH Aachen, now KIT), for his trust and the opportunity to be part of the iLoPS project. Thank you very much for the support and countless fruitful discussions during my PhD work and for the many reviews of my manuscripts. I really enjoyed our brainstorming sessions immensely. It is important for me to note that he made me learn a great deal about myself during this PhD project, which positively fueled my personal growth. Beyond the academic guidance throughout this time, this was of priceless value to me. Janos Urai is kindly acknowledged for being my second referee. I would also like to thank my colleagues, Benjamin Busch (former RWTH Aachen, now KIT) and Ulrike Hilse (former RWTH), for the many and long discussions about a multitude of scientific topics. A special thanks belongs to all the enthusiastic students involved in the Upper Carboniferous project. Teamwork is the key to success, and I highly appreciate the participation of every single student in the form of being student research assistants and/or carrying out (under-)graduate projects. Thereby I would like to especially acknowledge Ivy Becker, Daniel Bücken, Steffen Fündgens, Mareen Höhne, Markus König, Yasar Manß, Simon Schröer and Philip Steindorf, all of whom significantly contributed to this project. I am proud that Ivy and Yasar (former RWTH Aachen, now KIT) successfully finished their graduate projects and are now continuing their academic careers, themselves becoming PhD students. Bettina Leesmeister (RWTH Aachen) demonstrated how to manage bureaucratic work and had solutions for almost any problem. Thanks, as well, for having such open ears in many situations.

I am thankful to Tom Derichs and Werner Kraus (RWTH Aachen) who provided essential help and knowledge pertaining to sample preparation – good samples are fundamental to any scientific research. Many thanks to Pieter Bertier and Helge Stanjek (RWTH Aachen) for their fruitful discussions of clay mineralogy, diagenesis and XRD measurements. A special thanks belongs to Dennis Künkel (RWTH Aachen) for his essential support with XRD measurements and with clay mineralogical sample preparation. Alexandra Amann, Bernhard Krooss, Reinhard Fink (RWTH Aachen) and Norbert Schleifer (Wintershall Holding Germany GmbH) are thanked for helpful advice and for providing practical knowledge and support for the petrophysical measurements. The development of a new hydrostatic flow cell for permeability measurements was made possible by Alexandra Amann and Bernhard Krooss and relied on their many years of practical and theoretical knowledge. The implementation of the core research facility, with their help, was very interesting to me and it was amazing to learn how much work and patience is necessary. Many thanks, as well, to Ralf Littke, Jan Schwarzbauer and Alexander Stock (RWTH Aachen) for providing access to the vitrinite reflectance and TOC measurement facilities.

Norbert Klitzsch and Jan Niederau (RWTH Aachen) are recognized for the many interesting scientific conversations we had. Thanks are definitely warranted with André Hellmann (RWTH Aachen) for sharing his broad knowledge of any topic related to mineralogy, ores and economic geology. I am sincerely grateful to my co-authors for their help with improving my manuscripts and for providing valuable scientific insights. Further, I would like to thank Volker Lüders (GFZ Potsdam) for introducing me to the world of fluid inclusion studies, and I am also thankful to Klaus Wemmer (GAU Göttingen) for carrying out the K-Ar age determinations.

I would like to especially acknowledge Stefan Back (RWTH Aachen) for the many conversations. Your positive attitude, enthusiasm and neutral perspective really aided me in keeping on track and motivated during the last phase of my PhD – thank you! Hopefully, there will be occasions for whisky tastings in the future! Thanks to Kathrin Heinzmann (RWTH Aachen), too, for always having open ears and advice. I am also thankful to Marc David Joisten for personal conversations.

I wish to thank Wintershall Holding GmbH for funding this study. Bastian Koehrer is thankfully acknowledged for his countless hours reviewing written manuscripts and introducing me to Wintershall workflows and staff. On the Wintershall side of this project, I am also grateful to Dirk Adelmann, Maite de Medeiros, Norbert Schleifer, Philipp Antrett, Peter Süss, Gregor Hollmann, Dieter Kaufmann, Harald Karg and Wolf-Dieter Karnin for the various fruitful discussions. Thank you very much for your support, sharing valuable knowledge and supplying interesting insights from an industry perspective.

I want to express gratitude to Duncan McLean (MB Stratigraphy Ltd) for introducing me to the world of Carboniferous palynology and the very interesting field trip within the scope of this very special scientific topic. Angelika Leipner (Museum am Schölerberg) is recognized for the interesting debates and sharing knowledge during the several fieldwork visits to the Piesberg quarry.

Finally, I want to thank my family, Luisa and my friends for their support, encouragement and simply for being there for me under any circumstances. You and your immeasurable support made this work possible. I know that this time, with my ups and downs, was not just hard to me… sometimes, you had to be more than patient and indulgent with me! Thank you!

Abstract

Upper Carboniferous sandstones are one of the most important tight gas reservoirs in Central Europe. Data are derived in a kilometer-scale Upper Carboniferous reservoir outcrop analog (Piesberg quarry) in the Lower Saxony Basin, NW Germany. This field-based study focused on the diagenetic control on spatial reservoir quality distribution. Geothermometers were used to characterize a fault-related thermal anomaly in this reservoir-scale outcrop analog. A prototype workflow based on terrestrial laser scanning (t-LiDAR) is presented, which allowed for the automated detection and analysis of fractures.

The investigated outcrop consists of fluvial fining-upward cycles, which originate from a braided river dominated depositional environment. Westphalian C/D stratigraphy, sedimentary thicknesses and exposed fault orientations (NNW-SSE and W-E) reflect tight gas reservoir properties in the region further north. Diagenetic investigations revealed an early loss of primary porosity by pseudomatrix formation. Present-day porosity (7 % on average) and matrix permeability (0.0003 mD on average) reflect a high-temperature overprint during burial. The entire remaining pore space is occluded with authigenic minerals, predominantly quartz and illite. This reduces reservoir quality and excludes exposed rocks as tight gas targets. The correlation of petrographic and petrophysical data show that expected facies-related reservoir quality trends were overprinted by high-temperature diagenesis. The present day secondary porosity reflects the telogenetic dissolution of mesogenetic ankerite cements and unstable alumosilicates. Faults are associated with both sealed and partially sealed veins near the faults. Around the W-E striking faults, dissolution is higher in leached sandstones with matrix porosities of up to 26.3 % and matrix permeabilities up to 105 mD. The increased dissolution of ankerite and lithic fragments near the faults indicates focused fluid flow. However, the telogenetic origin cannot be ruled out. A variety of geothermometers (chlorite thermometry, fluid inclusion microthermometry and vitrinite reflection measurements) were employed to characterize the thermal anomaly in the studied reservoir outcrop analog, which is assumed responsible for high temperatures of circa 300°C, deteriorating the reservoir quality entirely. The tight gas siliciclastics were overprinted with temperatures approximately 90 – 120° C higher compared to outcropping rocks of a similar stratigraphic position some 15 km to the west. The local temperature increase can be explained by circulating hydrothermal fluids along the fault damage zone of a large NNW-SSE striking fault with a displacement of up to 600 m in the east of the quarry, laterally heating up the entire exposed tight gas sandstones. The km-scale lateral extent of this fault-bound thermal anomaly is evidenced by vitrinite reflectance measurements of meta-anthracite coals ($VR_{rot} \sim 4.66$) and the temperature-related diagenetic overprint. Data suggest that this thermal event and associated highest coalification was reached prior to peak subsidence during the Late Jurassic rifting (162 Ma) based on the K-Ar dating of the < 2 μm fraction of the tight gas sandstones. Related stable isotope data from fluid inclusions, hosted in a first fracture filling quartz generation (T ~ 250°C), close to the lithostatic fluid pressure (P ~ 1000 bars), together with authigenic chlorite growth in mineralized extension fractures, demonstrate that coalification was not subject to significant changes during ongoing burial. This is further evidenced by the biaxial reflectance anisotropy of meta-anthracite coals. A second event of quartz vein formation took place at lower temperatures (T ~ 180°C) at lower (hydrostatic) pressure conditions (P ~ 400 bars) and can be related to basin inversion. This second quartz generation might be associated with a second event of illite growth and K-Ar ages of 96.5 – 106.7 Ma derived from the < 0.2 μm fraction of the tight gas sandstones.

Understanding natural fracture networks in the subsurface is highly challenging, as direct 1D borehole data are unable to reflect their spatial complexity, and 3D seismic data are limited in spatial resolution to resolve individual meter-scale fractures. The workflow allows the t-LiDAR data to be integrated into conventional reservoir modeling software for characterizing natural fracture networks with regard to orientation and spatial distribution. The analysis outlines the lateral re-orientation of fractures from a WSW-ENE strike, near a normal fault with approximately 600 m displacement, towards a W-E strike away from the fault. Fracture corridors, 10 to 20 m wide, are present in unfaulted rocks with an average fracture density of 3.4 – 3.9 m^{-1}. A reservoir-scale digital outcrop model was constructed as a basis for data integration. The fracture detection and analysis serve as input for a stochastically-modeled discrete fracture network (DFN), demonstrating the transferability of the derived data into standard hydrocarbon exploration and production industry approaches. The presented t-LiDAR workflow provides a powerful tool for quantitative spatial analysis of outcrop analogs, in terms of natural fracture network characterization, and enriches classical outcrop investigation techniques.

The results of this work demonstrate both the transferability and limits of outcrop analog studies with respect to actual subsurface reservoirs of the greater area. Whereas the investigated outcrop forms a suitable analog with respect to sedimentological, stratigraphic and structural inventory, actual reservoirs at depth generally lack telogenetic influences, altering absolute reservoir quality values at the surface. This study shows the exploration risk of fault-bound thermal anomalies by entirely deteriorating the reservoir quality of tight gas sandstones with respect to porosity and permeability based on cementation with temperature-related authigenic cements. It documents that peak temperatures are not necessarily associated with peak subsidence. Consequently, these phenomena need to be considered in petroleum system models to avoid, for example, overestimates of burial depth and reservoir quality. The outlined workflow based on t-LiDAR may contribute to a better application of digital outcrop analog data into naturally fractured reservoirs at the subsurface, reducing uncertainties in the characterization of this reservoir type at depth.

Kurzfassung

Oberkarbonische Sandsteine gehören zu den wichtigsten Tight-Gas-Reservoiren in Mitteleuropa. Die präsentierten Daten wurden in einem oberkarbonischen Aufschlussanalog im Kilometermaßstab (Piesberg Steinbruch) in NW-Deutschland erhoben. Die auf Feldarbeit gestützte Studie beschäftigt sich mit der diagenetischen Kontrolle auf die räumliche Verteilung der Reservoirqualität. Eine störungsgebundene thermische Anomalie wurde mit Geothermometern charakterisiert. Es wird ein Arbeitsablauf vorgestellt, welcher die automatisierte Erkennung und Analyse von Trennflächengefügen, basierend auf terrestrischem Laserscanning (t-LiDAR), ermöglicht.

Der Aufschluss besteht aus fluviatilen, gradierten Zyklen, welche in einem von verflochtenen Flusssystemen dominiertem Ablagerungsraum entstanden. Das Westphal C/D, die sedimentäre Mächtigkeit und die aufgeschlossenen Störungsorientierungen (NNW-SSE und W-E) spiegeln Tight-Gas-Reservoir-Eigenschaften im Norden der Region wieder. Die frühdiagenetische Entwicklung einer Pseudomatrix verschließt primären Porenraum. Die heutige Porosität (Ø ~ 7%) und Permeabilität (Ø ~ 0.0003 mD) sind hauptsächlich das Resultat eines Hochtemperaturereignisses während der Versenkungsgeschichte. Der noch bestehende Porenraum wurde dabei mit authigenen Mineralen verschlossen, hauptsächlich durch Quarz und Illit. Die sekundäre Porosität repräsentiert des Weiteren die telogenetische Lösung von mesogenetischem Ankeritzement und unstabilen Aluminosilikaten. Störungen sind mit teilweise- und komplett zementierten Klüften assoziiert. Lösungsprozesse sind um W-E streichenden Störungen stärker ausgeprägt, wodurch eine Porosität von bis zu 26.3 % und eine Permeabilität von bis zu 105 mD erreicht werden. Die erhöhte Lösung von Ankerit und lithischen Fragmenten kann mit störungsgebundenen Fluidfluss erklärt werden. Allerdings kann dabei ein telogenetischer Ursprung nicht ausgeschlossen werden. Geothermometer (Chlorite, Fluideinschlüsse, Vitrinitreflexion) wurden für die Charakterisierung einer thermischen Anomalie im Reservoiranalog genutzt. Die untersuchten Sandsteine waren Temperaturen von ca. 300°C exponiert, wodurch die Qualität des Reservoirs drastisch verschlechtert wurde. Im Vergleich mit aufgeschlossenen Sandsteinen der gleichen Stratigraphie, ca. 15 km westlich, wurden die Tight-Gas-Siliziklastika ca. 90 – 120°C höheren Temperaturen ausgesetzt. Dieses lokale Phänomen kann mit hydrothermalen Fluiden erklärt werden, die entlang der Bruchzone einer großen NNW-SSE streichenden Störung, mit einem Versatz von bis zu 600 m, zirkulieren und die Formation lateral aufheizen. Die Vitrinitreflexion (VR_{rot}, Ø ~ 4.66 %) der Metaanthrazite und die temperaturbedingte diagenetische Modifikation des Untersuchungsgebietes weist den Kilometermaßstab dieser thermischen Anomalie nach. Basierend auf der K-Ar Datierung der < 2 µm Fraktion von den Sandsteinen ist das Temperaturereignis und die damit verbundene Maturität der Kohle bereits vor der tiefsten Versenkung im Oberjura (162 Ma) erreicht. Stabile Isotopendaten von Fluideinschlüssen einer ersten Quarzgeneration (T ~ 250°C, lithostatischer Fluiddruck von ~ 1000 bar) von Extensionskluftfüllungen - assoziiert mit authigenem Chlorit Wachstum - implizieren, dass sich der Inkohlungsgrad im weiteren Verlauf der Versenkung nicht signifikant änderte. Die biaxiale Reflektions-Anisotropie der Metaanthrazite unterstützt diese Beobachtung. Eine zweite kluftfüllende Quarzgeneration (T ~ 180°C) wird der Beckeninversionsphase zugeordnet, was mit einer zweiten Wachstumsphase von Illit und K-Ar Altern von 96.5 – 106.7 Ma (< 0.2 µm Fraktion der Sandsteine) assoziiert werden könnte. Das Verständnis von untertägigen, natürlichen Kluftnetzwerken ist herausfordernd, weil 1D Bohrlochdaten die räumliche Heterogenität nicht wiederspiegeln und 3D seismische Daten keine Brüche oder Störungen im Metermaßstab auflösen. Der in dieser Studie vorgestellte Arbeitsablauf ermöglicht die Integration von t-LiDAR Daten in konventionelle Reservoir-Modellierungssoftware für die Charakterisierung von natürlichen Kluftnetzwerken bezüglich Orientierung und räumlicher Verteilung. Die automatisierte Analyse verdeutlicht eine laterale Reorientierung der Klüfte von einem WSW-ENE Streichen, in der Nähe von einer Störung mit bis zu 600 m Versatz, zu einem W-E Streichen, distal von der Störung. Zonen mit einer erhöhten Kluftdichte (Ø = 3.4 – 3.9 m^{-1}) in ungestörten Sandsteinen sind 10 bis 20 m breit. Ein digitales 3D Aufschlussmodell dient als Basis für die Integration von einem stochastisch modellierten, diskreten Bruchnetzwerk (DFN), basierend auf der automatisierten Trennflächenerkennung und Analyse. Dies demonstriert die Nutzbarkeit von den erhobenen Daten in Standardanwendungen der E&P Industrie.

Diese Studie zeigt sowohl die Übertragbarkeit als auch die Einschränkungen von Aufschlussanalogen in Bezug auf untertägige Reservoire der Region. Der untersuchte Steinbruch ist ein geeignetes Analog in Bezug auf Sedimentologie, Stratigraphie und strukturellem Inventar. Die Qualität des Reservoiranalogs wird maßgeblich von telogenetischen Prozessen beeinflusst, welche untertägig nicht vorhanden sind. Die Studie verdeutlich das Explorationsrisiko von störungsgebundenen, thermischen Anomalien und der damit einhergehenden Verschlechterung der Reservoirqualität durch assoziierte Zementation mit authigenen Mineralphasen. Es wird gezeigt, dass die höchste Temperatur nicht notwendigerweise mit der tiefsten Versenkung zu korrelieren ist. Lokale Hochtemperaturphänomene müssen folglich in der Modellierung von Kohlenwasserstoffsystemen oder der Vorhersage von Reservoirqualitäten von Tight-Gas-Sandsteinen an ähnlichen strukturellen Positionen berücksichtigt werden, um z.B. Versenkungsteufen nicht zu überschätzen. Der präsentierte t-LiDAR Arbeitsablauf ist ein vielversprechendes Werkzeug für die quantitative, räumliche Analyse von Aufschlussanalogen in Bezug auf die Charakterisierung von natürlichen Kluftnetzwerken und bereichert die klassische Arbeit im Aufschluss. Diese Studie kann dazu beitragen, die Anwendung von digitalen Aufschlussanalogdaten für natürlich zerklüftete, untertägige Reservoire zu verbessen und damit die Unsicherheit der Charakterisierung von diesen Reservoirtypen zu reduzieren.

Contents

Declaration of originality

Chapter 2: As first author, Patrick Wüstefeld performed the data acquisition as well as sample analysis and evaluation. Simon Schröer, Markus König and Yasar Manß conducted the analyses of subsets of samples. This was supervised and reviewed by the author. Patrick Wüstefeld incorporated, extended and re-evaluated a subset of data prepared by Philip Steindorf and Mareen Höhne within their graduate project. A subsurface petrophysical dataset was provided by Wintershall Holding GmbH Germany. Assistance during the XRD analyses from Pieter Bertier is highly acknowledged. For the TOC measurements, Jan Schwarzbauer engaged in helpful discussions. Patrick Wüstefeld wrote the entire article with input and support from Christoph Hilgers, Ulrike Hilse, Bastian Koehrer and Dirk Adelmann.

Chapter 3: As first author, Patrick Wüstefeld performed the field sampling, data acquisition, sample analysis and evaluation. Volker Lüders conducted the fluid inclusion analyses and Klaus Wemmer carried out the K-Ar age determination of subsets of samples. Their aid in the data evaluation is highly appreciated. Petrohpysical analysis and evaluation of subsets of samples (Woitzel quarry) was performed by Ivy Becker as part of her graduate project, which was supervised and reviewed by the author. Patrick Wüstefeld wrote the entire article with input and feedback from Christoph Hilgers, Volker Lüders, Ulrike Hilse, Klaus Wemmer and Bastian Koehrer.

Chapter 4: As first author, Patrick Wüstefeld performed all data acquisition, analysis and evaluation. Steffen Fündgens developed the algorithm to automatically identify discontinuity surfaces from t-LiDAR point cloud data as part of his undergraduate project. Steffen Fündgens was supervised by Patrick Wüstefeld, Dominik Sibbing and Leif Kobbelt. The author developed the workflow to apply the algorithm to digital outcrop data and to integrate the data in reservoir modeling software. He also reviewed and evaluated the processed data. Patrick Wüstefeld wrote the entire article with input and support from Christoph Hilgers, Maite de Medeiros, Bastian Koehrer, Dominik Sibbing and Leif Kobbelt.

1 Introduction

1.1 Rationale

Low-permeability tight gas sandstones form unconventional hydrocarbon reserves and became progressively more important for gas exploration activities over the last decades (e.g. Law et al., 1989; Leeder and Hardman, 1990; Meckel et al., 1992; Besly, 1998; Law et al., 1998; Kombrink et al., 2010) because conventional reservoirs are becoming increasingly scarce. The term "tight gas reservoir" has ambiguous definitions and usually refers to basin-centered gas accumulations (e.g. Law, 2002; Law and Curtis, 2002). Within the scope of this study, this term refers to low-porosity/permeability systems, with permeabilities below 0.6 mD (conditions for North German reservoirs; q.v. Pusch et al., 2005; Häfner, 2006; Albrecht and Reitenbach, 2015). For economic development of this reservoir type, the application of reservoir stimulation techniques is usually inevitable (e.g. Holditch, 2006; Rezaee et al., 2012; Wimmers and Koehrer, 2014). The estimated tight gas resources are approximately 90 billion m^3 in Germany and 63.000 billion m^3 worldwide (BGR, 2016).

The Central European Basin System (e.g. Littke et al., 2008) hosts a large number of Paleozoic gas fields, reaching from the North Sea over the Netherlands and Germany to Poland (Glennie, 1986; e.g. Glennie, 1997; Gaupp et al., 2008; Littke et al., 2011). Tight gas sandstones of the Carboniferous host some of the most important tight gas reservoirs in Europe. The "sweet spot" prediction for tight reservoirs remains a major challenge and reservoir characteristics are still poorly understood (e.g. Cowan, 1989; Besly, 1998). The present day quality of a reservoir is controlled by the means of storage (porosity) and flow capacity (permeability). The productivity in tight formations can be very heterogeneous (e.g. Sakhaee-Pour and Bryant, 2014), which highlights the importance of identifying areas of increased reservoir quality. The control on reservoir properties, such as porosity and permeability, are manifold and mainly the result of the detrital and authigenic composition. The latter is strongly influenced by diagenetic and structural overprints during burial history. A careful assessment of controlling factors plays an important role in both the estimation of ultimate recoveries and the identification of risks in hydrocarbon exploration. Besides reservoir quality prediction in terms of matrix porosity and permeability, the understanding of natural facture networks is essential for tight formations. They tremendously affect flow properties and can be thus a major factor in exploitation success (Bourbiaux, 2010; Bahrami et al., 2011; Santos et al., 2015).

1.2 Objective

This study addresses the reservoir quality distribution and underlying processes of the Upper Carboniferous in the Lower Saxony Basin in NW Germany. It contributes to enhance the understanding of Upper Carboniferous hydrocarbon plays in the region. This field-based outcrop analog study focused on a reservoir-scale Upper Carboniferous outcrop at the southern margin of the Lower Saxony Basin. Analog studies overcome the limited resolution of seismic sections and the 1D information of limited core material, and is a widely applied approach to enhance the characterization of reservoirs (e.g. Miall, 1988b; Sharp et al., 2003; Pranter et al., 2007, 2014). The study was initiated to unravel spatial reservoir quality distribution for the tight gas reservoirs in NW Germany and addresses the general aspects relevant for tight gas reservoirs worldwide.

The key aim of the study was to depict the controlling factors of spatial and temporal reservoir quality distribution and evolution by integrating a variety of different geoscientific disciplines. As such, the study linked petrophysical datasets with the structural and diagenetic evolution throughout burial history. Furthermore, this work sought to determine the relationship of and interaction between sedimentology and structural and diagenetic evolution. It addresses the exploration risk of local thermal anomalies and links it to the temporal structural evolution. Besides the focus on the matrix properties of the tight gas siliciclastics, this study also had the objective of enhancing the understanding of the natural fracture network of the reservoir-scale outcrop analog.

1.3 Aims

The thesis focuses on the following aspects, which developed during the project period:

- The evaluation of the main control on reservoir properties and their evolution with special emphasis on the diagenetic overprint during burial history.
- The (temporal) impact of the structural evolution on diagenesis and eventual reservoir quality with a particular focus on the impact of local fault-bound thermal anomalies.
- The critical evaluation of outcrop analog studies and to what extent they are transferable to reservoirs at depth.
- The development of a workflow for the integration of high-resolution digital field mapping data based on terrestrial laser scanning (t-LiDAR) into conventional reservoir modeling software for characterizing natural fracture networks.

1.4 Geological Frame

The studied outcrop analog is an active quarry for Upper Carboniferous sandstones at the southwestern margin of the Lower Saxony Basin. The reservoir-scale Piesberg quarry near the town of Osnabrück has a dimension of approximately one km in the W-E direction and half a kilometer in N-S direction. The outcrop extends along the WNW-ESE striking Piesberg-Pyrmont axis (Drozdzewski, 1985; Drozdzewski et al., 2009) and is situated in the hanging wall of the Osning thrust (e.g. Baldschuhn and Kockel, 1999; Kley et al., 2008), which formed due to the Late Cretaceous inversion. In the east of the quarry, a NNW-SSE striking fault with a down-dip displacement of up to 600 m (Hinze, 1979; Baldschuhn et al., 2001) separates the Carboniferous siliciclastics from Mesozoic sediments. Clastic sedimentary rocks with intercalated coal seams dominate the Upper Carboniferous Westphalian C and D, which are exposed in the quarry. Rotliegend siliciclastics are known to be a feature of the central parts of the Lower Saxony Basin but are absent in the study area. Here, the Zechstein unconformably overlies the Upper Carboniferous (e.g. Voigt, 1960; Fiedler, 1984).

The basin evolution of the Lower Saxony Basin, belonging to the Central European Basin System, began with sedimentation in the foreland basin of the northwestwards prograding Variscan orogeny in the Late Carboniferous southeast of the study area (e.g. Ziegler, 1977; Glennie, 1986; Franke, 1995). In the Upper Carboniferous, the sedimentary environment changed from a marine to a (partially) terrestrial environment (e.g. Hedemann et al., 1984). The climate changed from humid to semi-arid in the Westphalian D in the region (Besly, 1998; Jones and Glover, 2005; Roscher and Schneider, 2006; Bertier et al., 2008). During this time, the area was dominated by large fluvial systems and the occurrence of swamp areas. The Central European Basin System evolution began in the Permian with rifting and/or thermal subsidence of the lithosphere (e.g. Bachmann and Grosse, 1989; van Wees et al., 2000; Adriasola-Muñoz, 2006). Likewise, the evolution of the Lower Saxony Basin was initiated with minor Permian subsidence associated with thermal lithosphere relaxation and sedimentation (e.g. Scheck and Bayer, 1999; van Wees et al., 2000). Rifting based on extension was the main control of Triassic subsidence, which continued until the Late Cretaceous (e.g. Brink et al., 1992; Senglaub et al., 2005; Lohr et al., 2007; Kley and Voigt, 2008). In the Late Mesozoic, the onset of the Alpine orogeny caused the re-configuration of the Central European Basin System and resulted in the inversion of several basins (e.g. Ziegler, 1982; Betz et al., 1987; Ziegler, 1987; Adriasola-Muñoz, 2006). In the Lower Saxony Basin, the inversion tectonics in the Late Cretaceous led to the exposure of Mesozoic sediments in many locations and, in rare cases, even Carboniferous sediments (e.g. Senglaub et al., 2005; Adriasola-Muñoz, 2006). This manifested, for example, through the uplifted Carboniferous blocks in the study area and the Ibbenbüren hard coal mining district (e.g. Drozdzewski et al., 2009). Inversion tectonics is one of the controlling factors that regulated the formation and modification of hydrocarbon traps in Carboniferous reservoirs in NW Germany and the North Sea (e.g. Kombrink et al., 2010). This is reflected, for example, by the prominent hydrocarbon reservoir blocks Husum-Schneeren and Rehden (e.g. Hollmann et al., 1998; Drozdzewski et al., 2009).

The southern margin of the Lower Saxony Basin is characterized by highest thermal maturity of coal measures (e.g. Teichmüller et al., 1984; Teichmüller, 1986; Petmecky et al., 1999). The reason for high maturity is discussed controversially. For many decades, a number of authors postulated a deep-seated pluton ("Bramscher Massif") being responsible for the anomalous maturity (e.g. Bartenstein et al., 1971; Stadler and Teichmüller, 1971; Teichmüller et al., 1984; Betz et al., 1987). This model was doubted at the beginning of this century and deep subsidence models were introduced in order to explain the highest temperatures at peak subsidence during the Late Cretaceous (e.g. Petmecky et al., 1999; Senglaub et al., 2006; Brink, 2013; Bruns et al., 2013).

1.5 Overview of the thesis

This doctoral thesis comprises five chapters. Besides the introductory chapter (1) and the final conclusive chapter (5), the three main chapters (2 – 4) are based on submitted manuscripts. They constitute self-contained research within the study area.

1.5.1 Diagenesis and petrophysical properties (Chapter 2)

This chapter deals with the controlling factors of spatial reservoir quality distribution derived in the Upper Carboniferous outcrop reservoir analog. The reservoir characteristics and their spatial heterogeneities of the tight fluvial sandstones are evaluated by integrating structural, stratigraphic and diagenetic controls. The chapter emphasizes diagenetic history and couples the same to petrophysical data. The work addresses the impact of the structural inventory on the diagenetic overprint. The temporal evolution of respective reservoir properties is essential, not least for critically assessing the transferability of outcrop data to actual reservoirs at depth. It documents the influence of high-temperature and telogenetic overprints on the present day reservoir properties. The chapter suitably also covers the limitations of outcrop analog studies with respect to actual subsurface reservoirs of the greater area.

1.5.2 Kilometer-scale fault-related thermal anomalies (Chapter 3)

This chapter details the anomalous thermal overprint of the tight gas reservoir analog, which has caused a complete deterioration of reservoir properties. The thermal anomaly is characterized by a variety of geothermometers and the temperature overprint is coupled to petrophysical data. The study establishes a local km-scale fault-bound thermal anomaly to be responsible for the temperature and associated diagenetic overprint. Furthermore, there is the careful addressing of the timing of such thermal events within the scope of basin and structural evolution. The results presented in this chapter effectively demonstrate the risk of hydrocarbon exploration of analogous structural positions as resulting diagenetic modifications have a primary impact on storage capacity and flow properties of the respective sandstone reservoir rocks. The chapter further highlights the importance of incorporating local thermal anomalies into petroleum system models.

1.5.3 New workflow to derive fracture statistics from t-LiDAR (Chapter 4)

This chapter discusses the characterization of natural fracture networks at the reservoir-scale by incorporating digital geology. For such tight formations, as investigated in the study area and respective subsurface reservoirs, the understanding of the natural fracture networks is fundamental because they can significantly improve flow properties. Within this chapter, a workflow is presented that allows for the characterization of fracture networks at the reservoir-scale based on t-LiDAR field data. The workflow is aligned with respect to data integration into modern petroleum industry workflows.

1.6 Parts of this thesis, which have been published

Articles:

Wüstefeld, P., Hilse, U., Koehrer, B., Adelmann, D., Hilgers, C., 2017. Critical evaluation of an Upper Carboniferous tight gas sandstone reservoir analog: Diagenesis and petrophysical aspects. Marine and Petroleum Geology 86, 689-710.

Wüstefeld, P., Hilse, U., Lüders, V., Wemmer, K., Koehrer, B., Hilgers, C., 2017. Kilometer-scale fault-related thermal anomalies in tight gas sandstones. Marine and Petroleum Geology 86, 288-303.

Submitted manuscripts:

Wüstefeld, P., de Medeiros, M., Koehrer, B., Sibbing, D., Kobbelt, L., Hilgers, C. Automated workflow to derive LIDAR fracture statistics for the DFN modelling of a tight gas sandstone reservoir analog. (submitted, in review).

Conference contributions:

Wüstefeld, P., de Medeiros, M., Koehrer, B., Sibbing, D., Kobbelt, L. & Hilgers, C. (2016). Automated workflow to derive LIDAR fracture statistics for the DFN modelling of a tight gas sandstone reservoir analog. EAGE 78th Conference and Exhibition, Vienna, Austria, 4 pp.

Becker, I., Wüstefeld, P., Hilse, U., Manß, Y., Koehrer, B., Hilgers, C. 2016 Reservoir heterogeneity of tight gas sandstones from observations of outcrop and subsurface data in the Lower Saxony Basin, Germany. DGMK/ÖGEW-Frühjahrstagung, Celle, Germany.

Wüstefeld, P., Hilse, U., Hilgers, C., Koehrer, B. (2016) Multiphase structural diagenetic controls on the spatial distribution of reservoir properties: A case study in an Upper Carboniferous tight gas sandstone reservoir analog. AAPG/SEG International Conference and Exhibition, Barcelona, Spain.

Becker, I., Wüstefeld, P., Hilgers, C., Koehrer, B., 2015. Sealing and leaching processes around normal faults in Upper Carboniferous tight gas sandstones from outcrop analog atudies. Geotectonic Research 97(1), 123-123.

Manß, Y., Wüstefeld, P., Hilse, U., Hilgers, C., Koehrer, B. (2015) Reservoir quality variations of Upper Carboniferous tight-gas sandstones on a 10's of kilometer-scale: Comparative and integrative core studies from two fields in NW-Germany. DGMK/ÖGEW-Frühjahrstagung 2015, Celle, Germany.

Wüstefeld, P., Hilgers, C., Koehrer, B. (2014) Structural diagenesis in an Upper Carboniferous Tight Gas Sandstones reservoir analog. AAPG International Conference and Exhibition 2014, Istanbul, Turkey.

Manß, Y., Wüstefeld, P., Hilgers, C., Koehrer, B., Bertier, P. (2014) Structural diagenesis of an Upper Carboniferous Tight Gas Sandstone Analog. 7th Mid-European Clay Conference 2014, Dresden, Germany.

Wüstefeld, P., Hilgers, C., Koehrer, B. (2014) Multi-scale reservoir heterogeneity and structural diagenesis in Upper Carboniferous tight gas sandstones: Insights from the Piesberg outcrop reservoir analog. DGMK/ÖGEW-Frühjahrstagung 2014, Celle, Germany.

Wüstefeld, P., Hilgers, C., Koehrer, B., Höhne, M., Steindorf, P., Schurk, K., Becker, S., Bertier, P., 2014. Reservoir heterogeneity in Upper Carboniferous tight gas sandstones: Lessons learned from an analog study. SPE/EAGE European Unconventional Resources Conference and Exhibition 2014, Vienna, Austria, SPE 167793, 10 pp.

2 Critical evaluation of an Upper Carboniferous tight gas sandstone reservoir analog: Diagenesis and petrophysical aspects

2.1 Abstract

Upper Carboniferous sandstones are one of the most important tight gas reservoirs in Central Europe. We present data from an outcrop reservoir analog (Piesberg Quarry) in the Lower Saxony Basin of Northern Germany. This field-based study focuses on the diagenetic control on spatial reservoir quality distribution.

The investigated outcrop consists of fluvial fining-upward cycles, which originate from a braided river dominated depositional environment. Westphalian C/D stratigraphy, sedimentary thicknesses and exposed fault orientations (NNW-SSE and W-E) reflect tight gas reservoir properties in the region further north. Diagenetic investigations revealed an early loss of primary porosity by pseudomatrix formation. Present day porosity (7 % on average) and matrix permeability (0.0003 mD on average) reflect a high-temperature overprint during burial. The entire remaining pore space is occluded with authigenic minerals, predominantly quartz and illite. This reduces reservoir quality and excludes exposed rocks as tight gas targets. The correlation of petrographic and petrophysical data show that expected facies-related reservoir quality trends were overprinted by high-temperature diagenesis. The present day secondary matrix porosity reflects the telogenetic dissolution of mesogenetic ankerite cements and unstable alumosilicates. Faults are associated with both sealed and partially sealed veins near the faults, indicating localized mass transport. Around W-E striking faults, dissolution is higher in leached sandstones with matrix porosities of up to 26.3 % and matrix permeabilities of up to 105 mD. The dissolution of ankerite and lithic fragments around the faults indicates focused fluid flow. However, a telogenetic origin cannot be ruled out and the increased reservoir properties in the fault corridors cannot be accounted for by reservoir-relevant processes based on the available data.

The results of this work demonstrate the limits of outcrop analog studies with respect to actual subsurface reservoirs of the greater area. Whereas the investigated outcrop forms a suitable analog with respect to sedimentological, stratigraphic and structural inventory, actual reservoirs at depth generally lack telogenetic influences. These alter absolute reservoir quality values at the surface. However, the temperature overprint and associated diagenetic modification, which caused the unusually low permeability in the studied outcrop, may pose a reservoir risk for tight gas exploration as a consequence of locally higher overburden or similar structural positions.

2.2 Introduction

The Central European Basin System comprises a large number of Paleozoic gas fields, from the North Sea over the Netherlands and Germany to Poland (Glennie, 1997; Gaupp et al., 2008; Littke et al., 2011). Carboniferous sediments comprise some of the most important tight gas reservoirs. They have become progressively more important for gas exploration (Leeder and Hardman, 1990; Besly, 1998; Petmecky et al., 1999; Kombrink et al., 2010). Due to complex tectonic and diagenetic histories, the reservoir characteristics of these formations are still poorly understood and "sweet spot" prediction is challenging (Cowan, 1989; Besly, 1998; Bertier et al., 2008).

The term "tight gas sandstone" is a commonly used ambiguous definition and in many cases equivalent to the terminology of basin-centered gas accumulations (Law, 2002). Spencer and Mast (1986) summarized several prominent examples of so-called tight gas sands in the United States. The definition of tight gas sandstones used for this outcrop analog study refers to low-porosity/permeability systems, which usually require the application of reservoir stimulation techniques (Rezaee et al., 2012; Wimmers and Koehrer, 2014). The estimation of ultimate recoveries of tight gas sandstone reservoirs is crucial for prospect evaluation. Predictions remain challenging and depend strongly on the pore structure and pore throats of the rock matrix (Bazin et al., 2010; Bahrami et al., 2011; Sakhaee-Pour and Bryant, 2014), which is mainly controlled by cement precipitation during burial history (Dutton et al., 1991). Diagenetic studies of permeable rocks are prevalent while less permeable formations have been hardly addressed (Stroker et al., 2013). Ajdukiewicz and Lander (2010) highlighted the demand for enhancing reservoir quality prediction by incorporating integrated diagenetic models that also consider petrophysical datasets. Moreover, productivity can significantly vary spatially in tight gas formations (Sakhaee-Pour and Bryant, 2014), underscoring the requirement to identify areas of enhanced reservoir quality and to integrate novel techniques to increase the chances of successful assessment of tight gas sandstone plays (Law, 2002). The characterization of fluvial reservoirs is even more complex, especially with respect to heterogeneities at different scales (Miall, 1988a; Sharp et al., 2003; Pranter et al., 2007). Analog studies overcome the limited resolution of seismic sections and the 1D information of limited core material, thus adding substantial value. Pranter et al. (2014) specifically noted the importance of assessing fluvial deposits in 3D and demonstrated the merit of integrating outcrop analog data with subsurface reservoir data.

In the work presented here, we focused on a large Upper Carboniferous outcrop in the Lower Saxony Basin. We derived the reservoir characteristics of tight fluvial sandstones and their heterogeneities in 3D by integrating structural, stratigraphic and diagenetic controls. Further, we demonstrate that despite excellent structural, lithofacies and stratigraphic, as well as apparent porosity correlations, the analog is not a full proxy of the subsurface reservoir. We showed the impact of a high-temperature overprint in tight gas sandstones on porosity and permeability and the associated diagenetic overprint.

2.3 Geological setting

The Piesberg quarry is located on the southwestern rim of the Lower Saxony Basin and is hosted in an area of active hard coal mining in the Upper Carboniferous Ibbenbüren district (Fig. 2.1). The large quarry (1025 m long, 620 m wide and 135 m deep) allows for the investigation of rock properties from the "seismic" down to the "sub-seismic" scale. It was chosen because of its similarity to reservoirs further north (Fig. 2.1) with respect to sedimentology, stratigraphy, structural fault patterns and average porosity. The area was dominated by large fluvial systems that interfered with the occurrence of swamp areas favored by the humid climate (Süss et al., 2000; Roscher and Schneider, 2006; Drozdzewski et al., 2008).

The fluvial cyclic deposits of the Westphalian C and D (Pennsylvanian) formations exposed in the Piesberg quarry are dominated by clastic sediments with intercalated coal seams (Fig. 2.2), and deposited as braided and meandering river deposits (David, 1987, 1990; Jones and Glover, 2005). Abundant coal seams and occasional intercalations of marine sediments characterize the Westphalian strata (Scheidt and Littke, 1989). Some of these marine deposits can be traced over hundreds of kilometers (Köwing and Rabitz, 2005; Drozdzewski et al., 2008). The base of the Westphalian D is set at the top of the „Zweibänke“ coal seam (Fig. 2.3) (Köwing and Rabitz, 2005). The climate in the Upper Carboniferous changed from humid to semi-arid within the Westphalian D in the region, resulting in local reddening of the uppermost section (Besly, 1998; Jones and Glover, 2005; Roscher and Schneider, 2006; Bertier et al., 2008). The Westphalian D strata of the Piesberg area are unconformably overlain by Upper Permian Copper Shale and Zechstein carbonate rocks (Fig. 2.2), the latter transformed to siderite and brown iron ore based on metasomatism (Voigt, 1960). Permian Rotliegend siliciclastics deposited in an arid environment are well-known from the central part of the Lower Saxony Basin but are absent in the study area (Voigt, 1960).

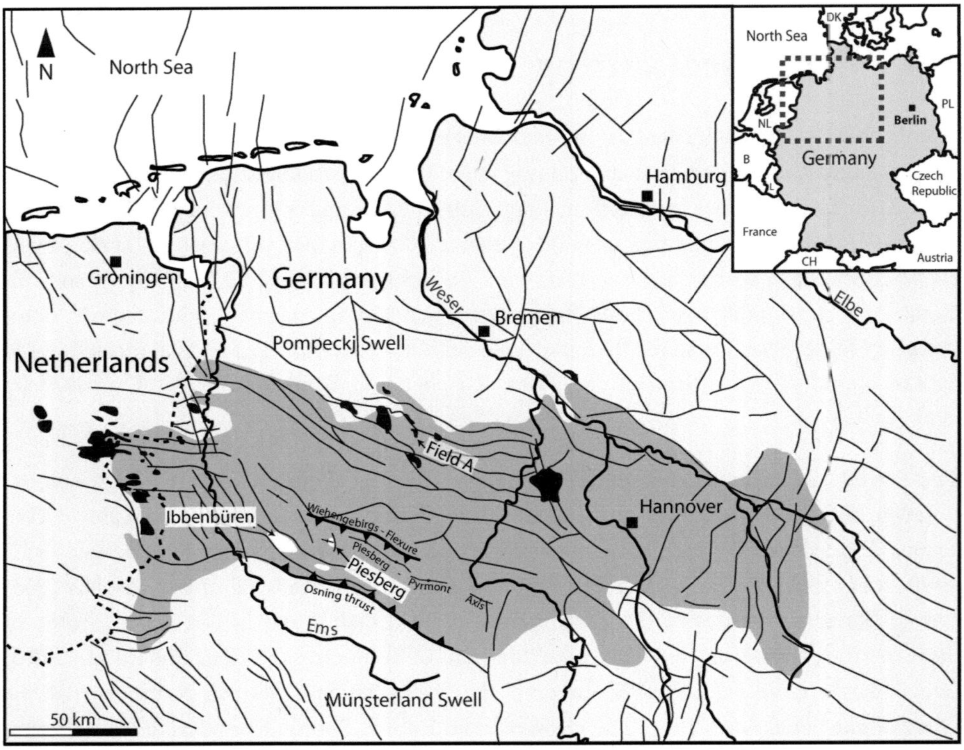

Figure 2.1: Carboniferous outcrops (white in dark grey area) at the southern rim of the Lower Saxony Basin (dark grey area, after Senglaub et al., 2006). The basement fault block pattern of the Lower Saxony Basin is based on Baldschuhn and Kockel (1999) and is extended with fault data from Kombrink et al. (2010) and Drozdzewski (1985). Black areas represent subsurface Carboniferous gas fields (after Kombrink et al., 2010).

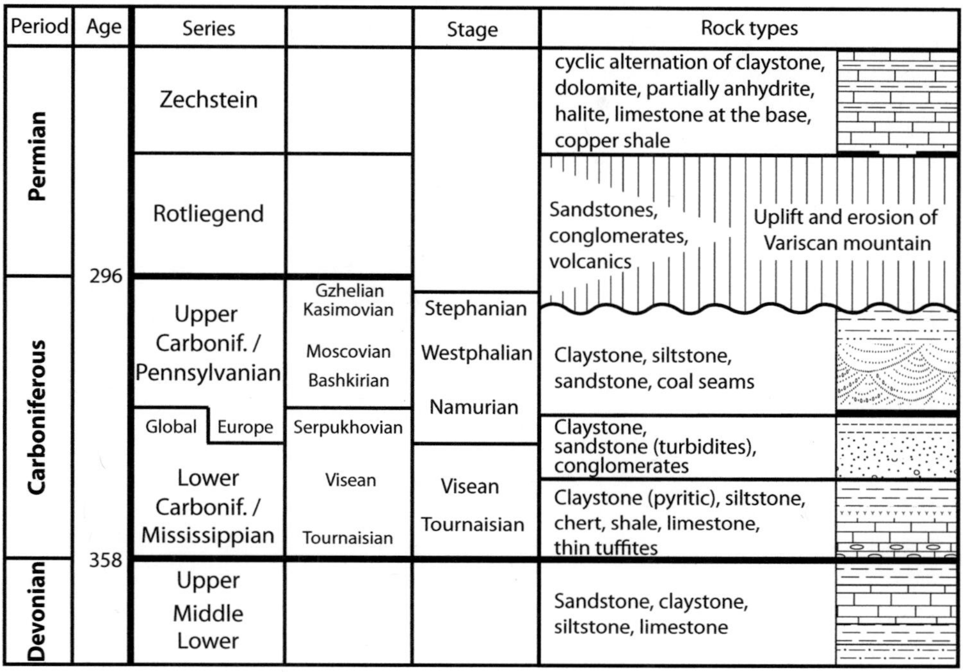

Figure 2.2: Stratigraphy of the Piesberg region (after Drozdzewski and Ribbert, 2003).

Structurally, the quarry belongs to the Variscan WNW-ESE striking Piesberg-Pyrmont axis (Fig. 2.1) (Drozdzewski, 1985; Drozdzewski et al., 2009). It forms an anticline with a wide interlimb angle that plunges gently to the west (Fig. 2.4). Folding was first explained by vertical movements (Hollmann et al., 1971) and later by dextral strike-slip on W-E directed faults (Drozdzewski, 1998). At the quarry's eastern side, a NNW-SSE striking fault with a down-dip displacement of up to 600 m (Fig. 2.4) (Hinze, 1979; Baldschuhn et al., 2001) separates the Upper Carboniferous from the Triassic Buntsandstein. Both fault orientations reflect the overall structural pattern in the region (Fig. 2.1). Late Cretaceous inversion tectonics induced an uplift of the Upper Carboniferous (Senglaub et al., 2005) and subsequent minor inversions led to the exposure of the rocks to the surface in the hanging wall of the Osning thrust (Fig. 2.1) (Baldschuhn and Kockel, 1999). The Late Cretaceous inversion is a main control for the formation or modification of hydrocarbon traps in Carboniferous reservoirs in NW Germany (Kombrink et al., 2010).

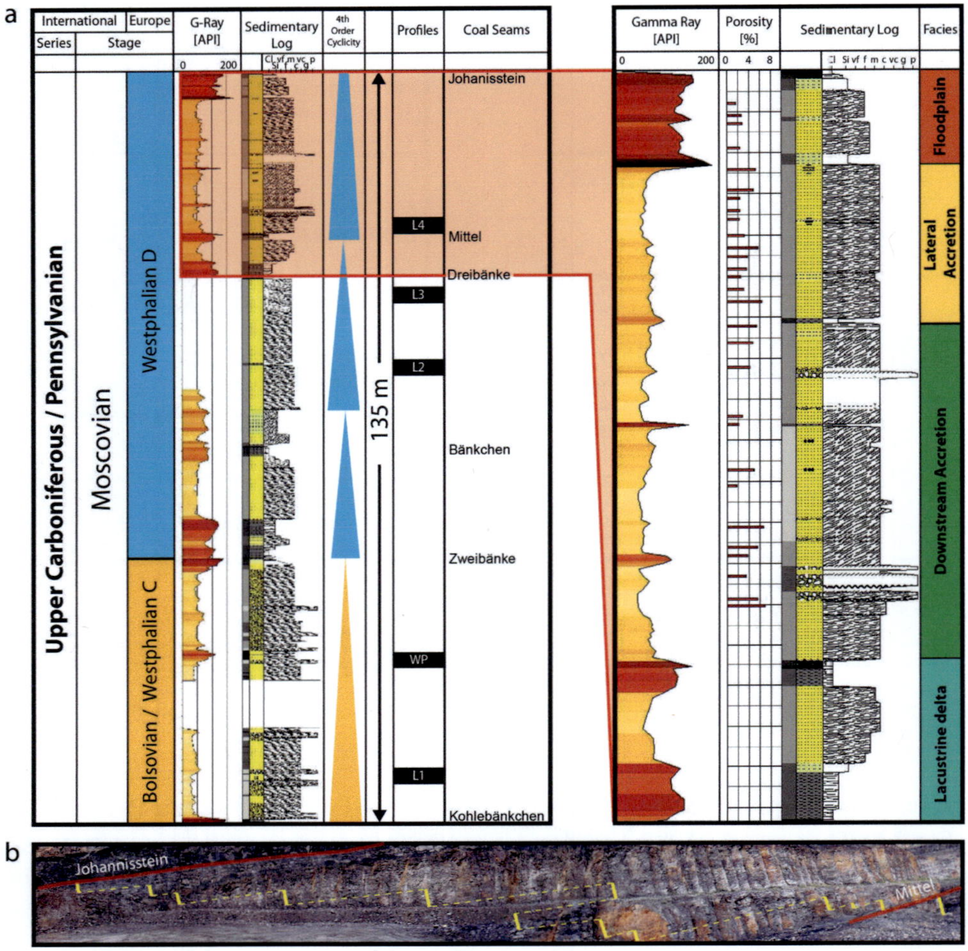

Figure 2.3: a) Stacked vertical sedimentological log covering 135 m of the accessible stratigraphy comprising the lithology as well as a gamma ray log (API units). Grain sizes follow the Wentworth (1922) classification (Cl = clay, Si = silt, vf = very fine sand, f = fine sand, m = medium sand, c = coarse sand, vc = very coarse sand, g = granule, p = pebble). Based on the logs, fourth-order stratigraphic cycles were interpreted. The black marks indicate the porosity wall panel (WP, Fig. 2.9) and the lateral porosity profiles (L1 – L4, Fig. 2.10). The magnification on the right shows the horizon between the „Dreibänke“ and „Johannisstein“ coal seams.
b) Location of stacked vertical sections (yellow) of the same horizon between the „Dreibänke“ and „Johannisstein“ coal seams. The section consists of individually logged subsections along the overall inaccessible 135 m high cliff.

The Lower Saxony Basin is characterized by high thermal maturity of organic matter in the south with thermal anomalies related to tectonic structures (Petmecky et al., 1999). The Piesberg coals feature the highest coalification rank of meta-anthracite stage coals with an R_0 of 4.8 and 4.9 (Bässler and Hoyer, 1971; Teichmüller et al., 1984; Teichmüller, 1986; Betzer et al., 2003). The high thermal maturity, with its peaks in the "Wiehengebirge" flexure zone (Fig. 2.1), was first related to a deep-seated intrusion (e.g. Bartenstein et al., 1971; Betz et al., 1987). More recently, it was explained with deep burial models down to at least 7 km (Senglaub et al., 2006; Bruns et al., 2013). The uplift during the Late Cretaceous inversion affected both tight gas reservoirs and the study area. While reservoirs are still at depths of 4 to 5 km (Baldschuhn and Kockel, 1999), the study area reached temperatures below approximately 60°C since the Paleocene and was finally uplifted in the Miocene (Senglaub et al., 2005).

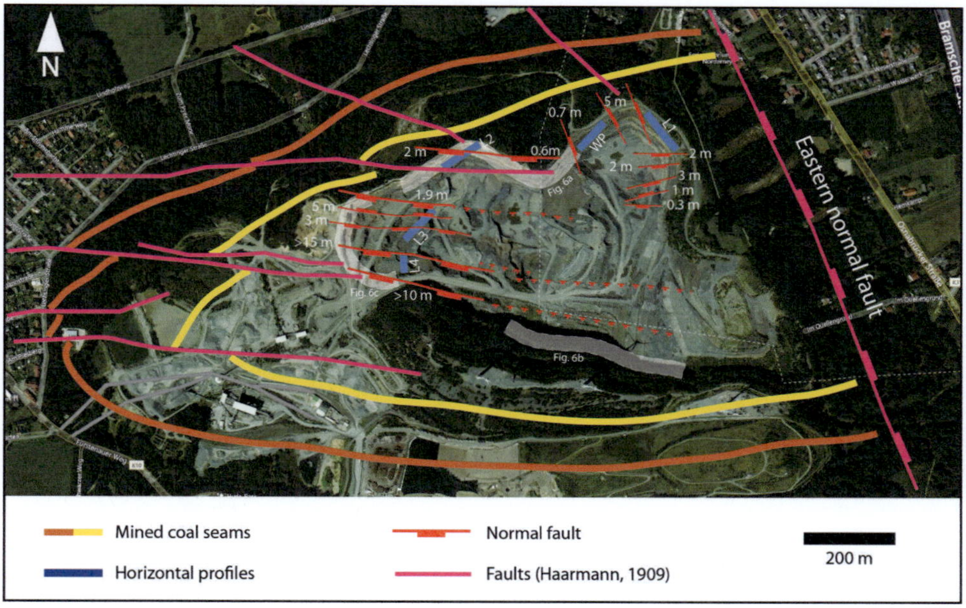

Figure 2.4: Aerial image of the Piesberg quarry (Bing Maps, 2015). Coal seams and normal faults (pink) as well as the major normal fault in the east are based on the map of Haarmann (1909). Faults (red) with indicated vertical offset represent faults observed in this study. The lateral profiles (L1 – L4, blue) as well as the wall panel (WP, blue) are depicted in Figs. 2.9 and 2.10.

2.4 Methodology

The entire exposed stratigraphic column was investigated in detail along a 135 m high vertical section with a focus on lithology and sedimentological structures. The vertical profile consisted of several stacked subsections because the cliff is inaccessible. However, the overall sedimentary thickness is constant with continuous coal seams across the quarry. Spectral gamma ray was measured with a handheld, auto-stabilizing RS-230 BGO Super-SPEC spectrometer with a 103 cm^3 bismuth germinate detector (Radiation Solutions Inc., Canada). Only fresh surfaces of the rock wall were measured with a counting time of three minutes. The yielded concentration analysis of potassium, thorium and uranium was converted to API units following Rider and Kennedy (2011):

$$API = K(\%)\times 16.32 + U(ppm)\times 8.09 + Th(pp)\times 3.93$$

Rock sampling along the vertical and horizontal sections required in situ drilling of plugs based on the rock hardness. In total, 228 one-inch plugs were drilled along the vertical and lateral profiles and used for petrophysical measurements (179 porosity and 40 permeability measurements). Additionally, there was one lateral profile within a single stratigraphic cycle based on 50 cutting samples (Wüstefeld et al., 2014).

Matrix porosity on plugs from the quarry material was determined with helium pycnometry by a non-commercial device as described in Ghanizadeh et al. (2014). The measuring circuit, including the sample and reference cell, was tempered to 30°C. The measurements were carried out on plugs dried at least overnight (105°C, vacuum). Porosities were derived from skeletal and bulk volumes as determined by geometrical calculations. For one lateral profile, the porosity was measured on cuttings (cf. Wüstefeld et al., 2014) where bulk density was determined according to the principle of Archimedes (Moussa, 1973).

Permeability of the plug samples from the quarry was measured by a steady-state technique using single-phase gas flow. The permeability was determined with Darcy's law (e.g. Ghanizadeh et al., 2013; Amann-Hildenbrand et al., 2016). A non-commercial isostatic flow cell, based on a hydrostatic pressure cell, was utilized to assess permeability of the very tight outcrop samples at confining pressures of 30 MPa and an inlet pressure of up to 10 MPa with helium as the permeant. A Klinkenberg correction was applied to all measured gas permeability to account for slip-flow effects.

In order to compare the porosity and permeability with the subsurface reservoir, the industry partner of this work provided data comprising 227 plug measurements. Porosity was determined by helium pycnometry and the Klinkenberg-corrected permeability measurements were performed with a commercial setup at 3 MPa confining pressure and air as permeant. The tight outcrop samples would not have been able to be measured by the same conventional technique at 3 MPa confining pressure because their rock permeability is below the lower detection limit.

Permeability of tight sandstones is dependent on the confining pressure (e.g. Amann-Hildenbrand et al., 2016). In order to account for stress dependency, a conversion of the subsurface dataset to 30 MPa confining pressure was applied following an equation proposed by David et al. (1994):

$$K = K_0 \times e^{-\gamma(P_{eff}-P_0)}$$

This formula describes permeability reduction as a consequence of increased confining pressures. The required sensitivity coefficient γ was derived from six plugs of the subsurface reservoir, which were re-measured at confining pressures of 10, 20, 40 and 60 MPa. K_0 is the determined permeability at the applied confining pressure P_0 (3 MPa). K is the converted permeability for the effective confining pressure P_{eff} (30 MPa). The differences in and resulting from the measurement techniques were neglected as this study sought to compare the ranges of permeability and not absolute values.

Petrographic investigation was carried out on 90 thin sections from selected plug caps, which were impregnated with blue-dyed or yellow fluorescent epoxy resin to highlight porosity. Several plugs were later used for petrophysical studies. Point-count analyses were conducted on 51 samples using a Pelcon Automatic Point Counter with 300 counts per thin section. The step size between each point was adjusted to maximum grain sizes (e.g. Howarth, 1998). Grain size and sorting analysis was performed by measuring the long axis of at least 100 grains per thin section on a grid adjusted to the maximum observable grain size to gain area-weighted results. The intergranular volume (IGV) was calculated by the sum of intergranular porosity, pore-filling cements and depositional matrix (e.g. Lundegard, 1992; Ehrenberg, 1995; Paxton et al., 2002). The estimation of the compactional porosity loss (COPL) and cementational porosity loss (CEPL) was based on formulas proposed by Ehrenberg (1995). For the calculations, an initial sandstone porosity of 40 % was assumed. The sorting was calculated from the grain size measurements (e.g. Tucker, 1991). Petrographic descriptions were supplemented with analyses of selected samples by scanning electron microscopy using a Zeiss ULTRA PLUS FE-SEM equipped with backscatter (BSE) and secondary electron detectors as well as an energy dispersive spectrometer system (EDX). The measurements were carried out with an accelerating voltage of 15 kV and a working distance of approximately 10 mm.

An electron microprobe (JEOL Superprobe JXA-8900R, measurement conditions: accelerating voltage of 15 kV, beam current of 24 mA, beam diameter of 10 µm, ZAF-correction) was used to assess chlorite elemental composition on polished thin sections.

The quantitative bulk mineralogy was determined by x-ray diffraction (XRD) with 18 samples. The sample material was crushed in a mortar to a grain size of 0.5 mm and then subsequently further milled with a McCrone Micronizing mill for 15 minutes with ethanol. 20% corundum was employed as an internal standard for improving measurement accuracy. The diffraction patterns from random powder compounds were analyzed, whereby care was taken to avoid preferential orientation of crystals. The measurements were carried out on a Bruker D8-Advance diffractometer using Cu-anodes equipped with a Sol_X detector. The quantitative phase analysis was conducted via Rietveld refinement utilizing *BGMN* software (Bergmann and Kleeberg, 1998) with customized clay mineral structure models (Ufer et al., 2008).

For the total organic carbon (TOC) measurements on the clay fraction, the < 2 µm fraction was separated from two selected samples with the Atterberg method after mechanical milling. Approximately 20 mg of sample material was employed for the TOC measurements with a LiquiTOC II device implemented with a solid matter module. A 5.31 % standard was used for calibrating the measurements. Samples were combusted between temperatures of 520°C and 1050°C. Measurements were performed with oxygen as the carrier gas and an infrared detector.

2.5 Results

2.5.1 Sedimentological logging

The quarry exposes grey fluvial sedimentary fining-upward cycles dominated by 4 to 40 m thick sandstone bodies with conglomeratic base grading upwards into 5 to 10 m thick siltstones and shales with intercalated coal seams (Fig. 2.3). The net sand volume of the formation is approximately 80 %. Sandbody geometries were previously documented by Jones and Glover (2005), who already highlighted the large lateral extent of the channel belts up to tens of kilometers wide for the study area, and hundreds of meters or more in width for individual channels. The compound sandbodies of the channel belts consist of deposits of individual amalgamating channels (Jones and Glover, 2005). Typically, their internal cross-beds feature different bar and bed forms (Fig. 2.5), which indicate downstream accretion dominated channel deposits at the base and lateral accretion dominated channel deposits at the upper part of the cycles (David, 1990; Jones and Glover, 2005). Erosive and irregular contacts between individual channel deposits (Fig. 2.5a) and conglomeratic lags (Fig. 2.5b) are common for downstream accretion dominated channel deposits. Conglomeratic lenses (Fig. 2.5c) occur at the base of the cyclic successions. Generally, mid-channel bar forms (Fig. 2.5a) are absent in the lateral accretion dominated channel deposits (Jones and Glover, 2005). The sandbody between the „Zweibänke“ and

„Bänkchen“ coal seams varies laterally in terms of thickness, ranging from 10 m on the northern quarry wall to 20 m on the southern quarry wall over a lateral distance of 200 m (Fig. 2.6 a,b). Within the overall fining-upwards cycles, the horizon between the „Dreibänke“ and „Mittel“ coal seams exposes a coarsening upward sequence in both lithology description and gamma ray log (Fig. 2.3).

Figure 2.5: a) Mid-channel bar forms and erosive contacts between individual channel deposits. Floodplain deposits overlie the channel sandbodies. b) Conglomeratic channel deposits exhibit quartz pebbles at the base. c) Conglomeratic lens. d) Lateral accretion deposits.

Figure 2.6 (Description on next page)

Figure 2.6 (Previous page): a) Northern quarry wall (cf. Fig. 2.4 for the location) featuring the top of the Westphalian C indicated by the „Zweibänke" coal seam and the cycles of the Westphalian D. The stacked vertical profile in Fig. 2.3b is marked by the red box. The distance between the „Bänkchen" and „Zweibänke" coal seams is approximately 10 m. b) Wall panel of the southern quarry wall (cf. Fig. 2.4 for the location) shows the „Zweibänke", „Bänkchen" and „Dreibänke" coal seams, which can be correlated across the W-E striking normal faults with the northern quarry wall (see Fig. 2.6 a,c). The distance between the „Bänkchen" and „Zweibänke" coal seams is approximately 20 m in this section. Note the variation in thickness of the sandbodies between both coal seams. The architecture of the channel belt deposits, with major erosional surfaces of individual channels, underscores the internal lateral variation of the sandstone bodies. c) View of the quarry in the western direction (cf. Fig. 2.4 for the location). Coal seams and faults are indicated. The „Zweibänke" coal seam delineates the Westphalian C/D boundary. The stacked vertical profile in Fig. 2.3b is marked by the red box. Note the truck for scale.

2.5.2 Structure

2.5.2.1 Faults

Two different fault directions strike NNW-SSE and W-E to WNW-ESE (Figs. 2.4, 2.6c). The most prominent normal fault, which is no longer exposed and locally referred to as "Ostrandverwerfung", strikes NNW-SSE and offsets the Paleozoic Westphalian D against the Lower Triassic by up to 600 m (Fig. 2.4) (Hinze, 1979; Baldschuhn et al., 2001). Several minor normal faults of similar strike and throws of up to 5 m are exposed on the northern and southern quarry wall (Fig. 2.4). The W-E to WNW-ESE striking normal faults have throws between 2 m and more than 15 m (Fig. 2.4). Some of these faults can be traced throughout the quarry (Fig. 2.4). All large faults feature clay smearing into the fault planes. Faulted sandstone beds show dilational jogs partially filled with euhedral quartz and locally with iron (oxide-)hydroxides. Indicators of displacement were found on a W-E striking fault, showing younger strike-slip-slickenlines truncating older dip-slip slickensides.

2.5.2.2 Veins and fractures

Subvertical quartz veins are oriented subparallel to the W-E striking faults. They are both fully and partially sealed and restricted to the vicinity of the faults with a larger offset. Around a fault with approximately 15 m offset, the veins occur in a 36 m wide corridor, their spacing increasing away from the fault. Quartz veins in open dilational jogs are locally covered by hematite crusts. Subvertical fracture sets post-date veins and strike NE-SW and WNW-ESE in the western part of the quarry (Fig. 2.7a). Near the major normal fault in the East (Fig. 2.4), the dominant fracture sets change orientation towards WSW-ENE and NNW-SSE (Fig. 2.7b).

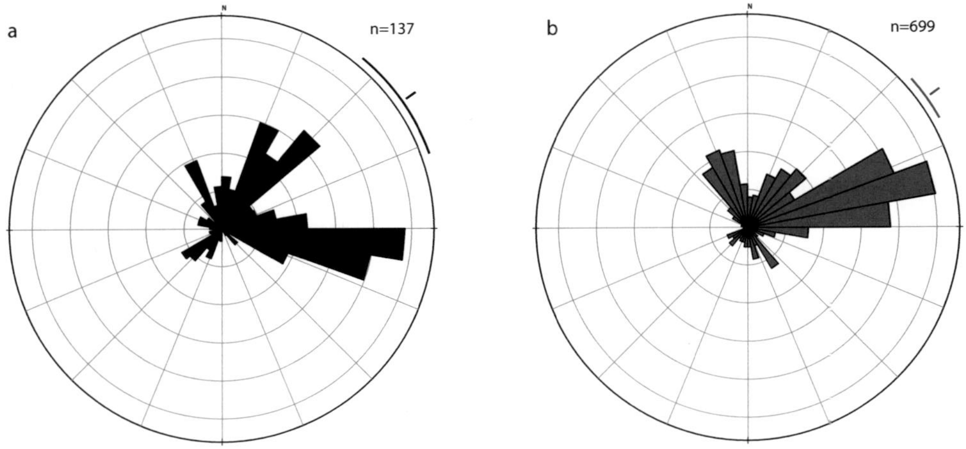

Figure 2.7: Rose diagrams (created with Openstereo software, Grohmann and Campanha, 2010) displaying the strike of fracture sets a) in the western part of the quarry and b) in the eastern quarry from two orthogonally oriented scanlines.

2.5.3 Porosity and permeability

Porosity of the investigated sandstones show an average (arithmetic mean) porosity of 7 % ranging from 0.5 to 26.3 % (Fig. 2.8). Mean (geometric) permeability of the samples is 0.0003 mD (at 30 MPa confining pressure) ranging between 40 nD to 105 mD (Fig. 2.8). Porosity and permeability exhibit a positive correlation (Fig. 2.8). The porosity in a vertical section of one fluvial fining-upward cycle ranges from 7.1 % in the sandstones to 1.7 % in the siltstones (Fig. 2.3a). However, the vertical profile highlights a significant intra-interval heterogeneity and variation. This is reflected in at least two trends of upward decreasing porosity (Fig. 2.3a). Porosities within a single medium-grained sandstone bed range between 4.5 to 8.7 % along a 56 m profile without consistent variation neither laterally nor vertically (Fig. 2.9). In the vicinity of normal faults (Figs. 2.4, 2.6c), matrix porosities can increase locally up to 25.5 % in lateral sampling profiles (Fig. 2.10) and permeabilities up to 105 mD (Fig. 2.8). Corridors of increased porosities can be 3 to 5 m wide in both directions across the fault (Fig. 2.9) (Wüstefeld et al., 2014) and are indicated by a whitish color variation. In other fault corridors, increased matrix porosities could not be verified (profile 4, Fig. 2.10). A light grey sample taken out of one such fault corridor features a matrix porosity of 26.3 %, while the same, but slightly darker, sandstone at an approximately 50 cm distance parallel to the fault plane has a matrix porosity of just 14.4 % (Fig. 2.11). In order to exclude the contribution of microfracturing to highly porous samples, we inspected thin sections taken perpendicular to the plug orientation, and no microfractures were observed. Furthermore, we tested a high-porosity sample at varying confining pressures up to 30 MPa. Porosities decreased from 28.6 % at 0.12 MPa to 27 % at 30 MPa and thus did not indicate a closure of potential microfractures within the high-porosity sample.

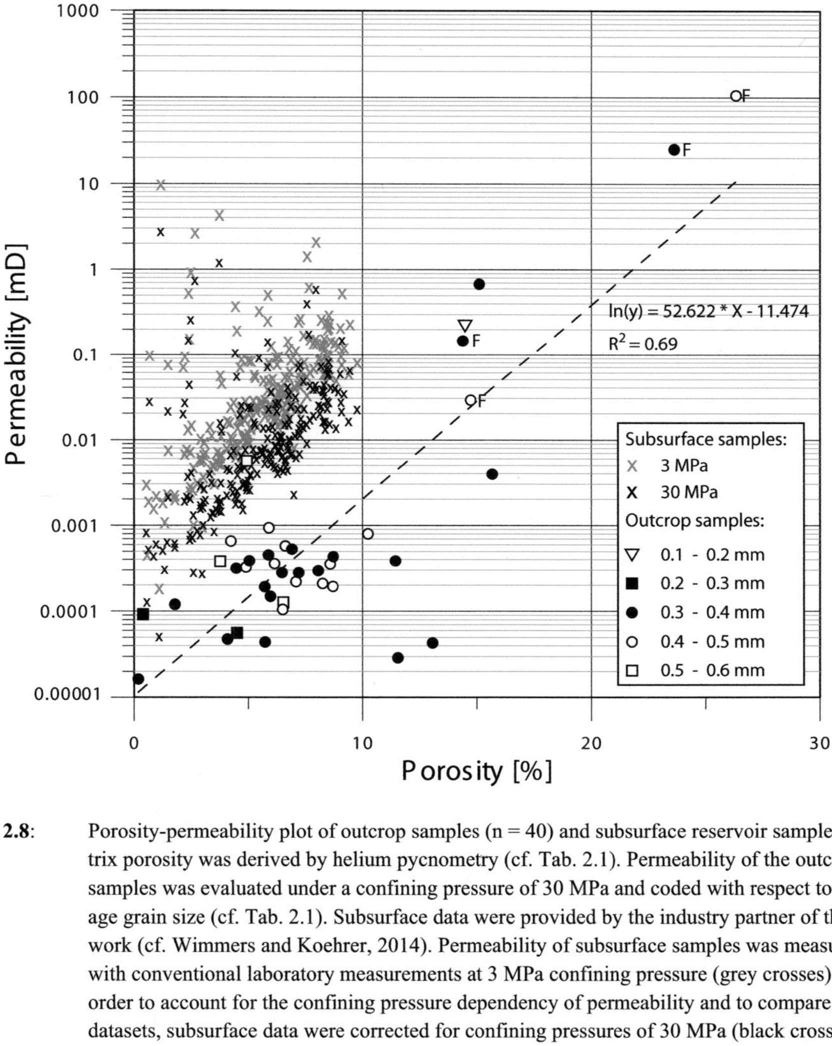

Figure 2.8: Porosity-permeability plot of outcrop samples (n = 40) and subsurface reservoir samples. Matrix porosity was derived by helium pycnometry (cf. Tab. 2.1). Permeability of the outcrop samples was evaluated under a confining pressure of 30 MPa and coded with respect to average grain size (cf. Tab. 2.1). Subsurface data were provided by the industry partner of this work (cf. Wimmers and Koehrer, 2014). Permeability of subsurface samples was measured with conventional laboratory measurements at 3 MPa confining pressure (grey crosses). In order to account for the confining pressure dependency of permeability and to compare both datasets, subsurface data were corrected for confining pressures of 30 MPa (black crosses). The matrix porosity of most outcrop analog samples and the subsurface reservoir are in the same range. The permeability of the outcrop analog is between one to two orders of magnitude smaller compared to the subsurface data. This is the result of a mesogenetic high-temperature overprint during burial and associated more intense authigenic cementation with predominantly quartz and illite. The present day porosity of the outcrop analog in turn is mainly a consequence of telogenetic dissolution processes. The leached rocks in the fault zones (outcrop samples marked with an ‘F’) have permeabilities up to five orders of magnitude higher than ‘regular’ sandstone samples and porosities up to 26.3 %.

Figure 2.9: Porosity wall panel (WP on Fig. 2.4), showing matrix porosity (measurements with helium pycnometry) within a medium-grained sandstone bed below the „Zweibänke" coal seam at a scale of 56 m with a sampling distance of 4 m laterally and 2 m vertically (Figs. 2.3a, 2.4). Each point represents a plug sample that was directly drilled out of the quarry wall. Average porosity is 5.9 % with a standard deviation of 1.1 %.

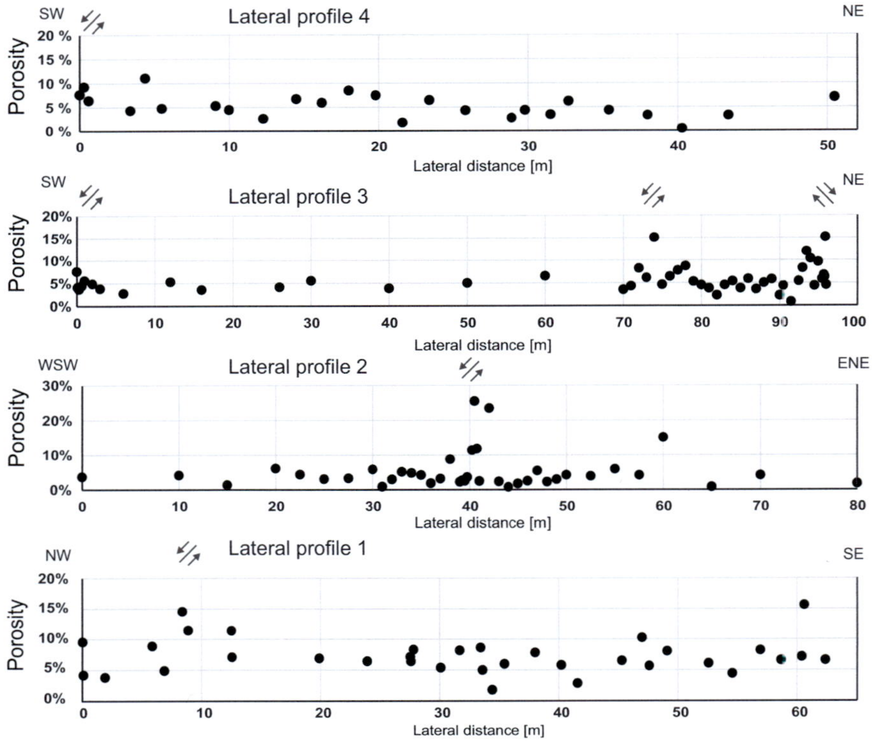

Figure 2.10: Porosity profiles restricted to single beds as indicated in the stratigraphic column (Figs. 2.3a, 2.4). Lateral distances range from 30 to 95 m. Position of faults are indicated in each profile. The lithology of lateral profile 1 consists of medium- to coarse-grained sandstones, which are pebbly in places. Lateral profiles 2 to 4 were sampled from medium-grained sandstones. Matrix porosity was determined by helium pycnometry.

Figure 2.11: Plug of extensively leached sandstone with light grey colors (left) has a matrix porosity of 26.3 % and a permeability of 105 mD (at 30 MPa confining pressure), while the slightly darker sandstone of the same bed has a porosity of 14.4 % and a permeability of 0.15 mD. Both samples were taken directly out of a fault zone of W-E striking faults with an offset > 15 m (Fig. 2.4) and less than half a meter between both sampling positions.

2.5.4 Sandstone petrology

The Westphalian C and D sandstones (Fig. 2.12) are classified as sublithic arenites with low volumes of feldspar grains (Fig. 2.13) based on point-count data (Tab. 2.1) following the classification of McBride (1963). The investigated sandstones are predominantly medium-grained (300 to 400 μm), moderately sorted sandstones (Tab. 2.1) with subangular to subrounded grain morphology.

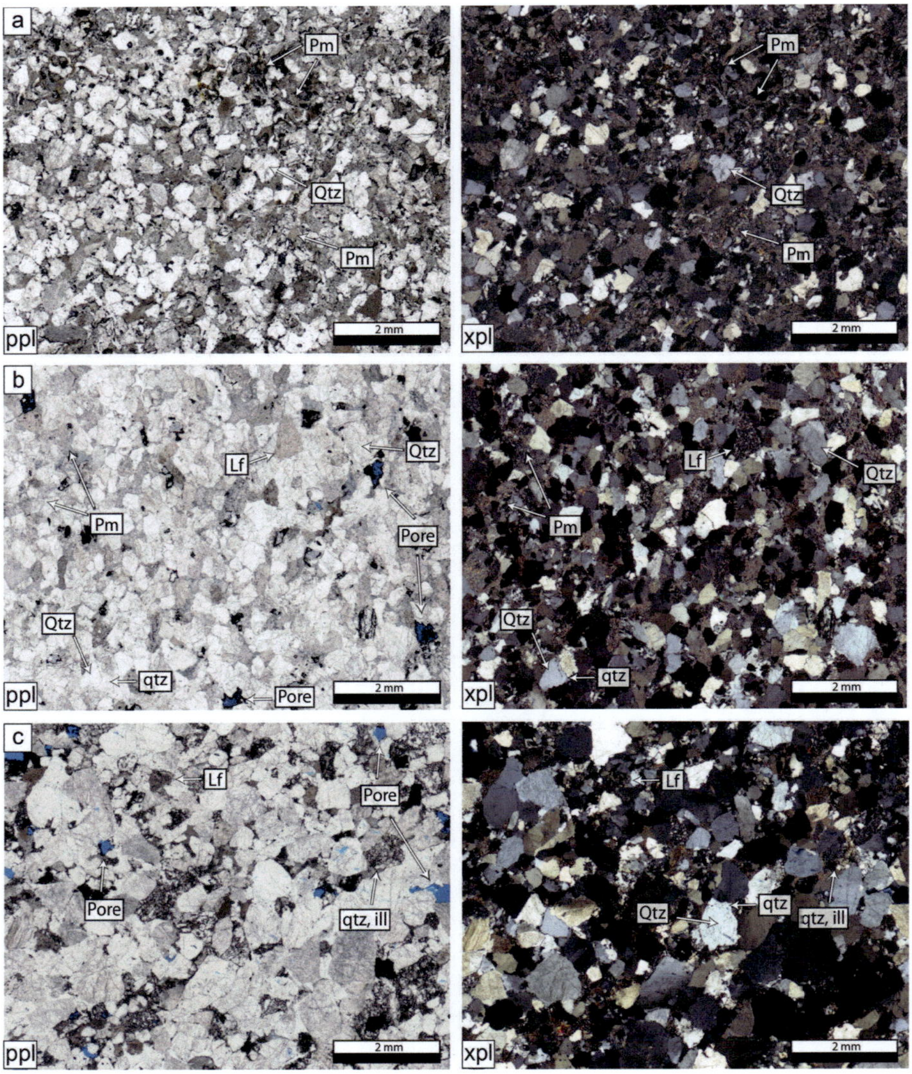

Figure 2.12: Plane-polarized (ppl) and crossed-polarized light (xpl) overview photomicrographs of the Westphalian C and D sandstones of the studied outcrop analog. a) Medium-grained sample SS16 (cf. Tab.1) with a large amount of pseudomatrix (Pm). The main detrital component is quartz (Qtz). The sample exhibits no optical porosity. b) Medium-grained sample MK12 comprises small amounts of pseudomatrix (Pm) besides detrital quartz (Qtz) and lithic fragments (LF) (cf. Tab. 2.1). Authigenic quartz (qtz) forms syntaxial overgrowths on detrital quartz grains (Qtz). Open pore space (pore, blue) has often a rhombic shape. c) Coarse-grained sample PB018 contains no pseudomatrix and has overall smaller amounts of lithic fragments as compared to the previous samples (cf. Tab.1). Besides authigenic quartz overgrowths on single detrital quartz grains (qtz), quartz-cemented meshwork illite (qtz, ill) can be observed (for details see also Fig. 2.14g, h). Open pore space (pore, blue) has often a rhombic shape.

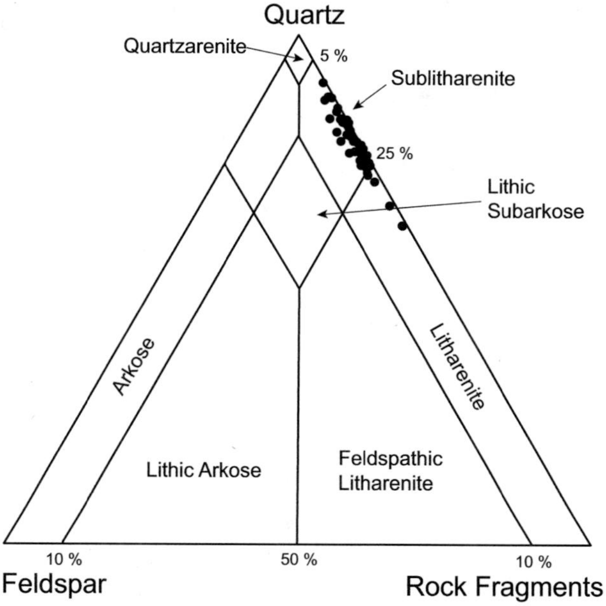

Figure 2.13: Rock classification of Piesberg samples based on point-counting in accordance with McBride (1963). Most samples can be classified as sublitharenite (n = 51).

2.5.4.1 Detrital mineralogy

The sandstones are mainly composed of quartz grains (Fig. 2.12) with an average volume of 57.5 % (mono- and polycrystalline, Tab. 2.1). Additional components include rock fragments as well as minor amounts of alkali feldspar (0.8 %) and muscovite (0.4 %) (Tab. 2.1). The feldspars are often strongly altered and replaced by authigenic minerals. The lithic fragment suite comprises shale and claystone fragments (2.3 %), such as clay rip-up and mud intraclasts, siliciclastic (feldspar-rich and sandstone) rock fragments (2.1 %) and metamorphic rock fragments (0.75 %). Ductile rock fragments, like shale, are often affected by deformation because of compaction (Fig. 2.14a). In addition to the alteration of lithic fragments (Fig. 2.14a), in most samples, high degrees of ductile deformation caused the formation of a pseudomatrix between the framework grains with a subsequent sericitization (Figs. 2.12, 2.14b). The criteria for classifying rock fragments as pseudomatrix implies that individual grains are not distinguishable anymore, but that the ductile deformation is visible through flow structures (e.g. Leeder and Hardman, 1990; Worden and Morad, 2003; Bertier et al., 2008). Although a detrital matrix cannot be ruled out, a differentiation is difficult due to the dominance of the altered pseudomatrix (Fig. 2.14b).

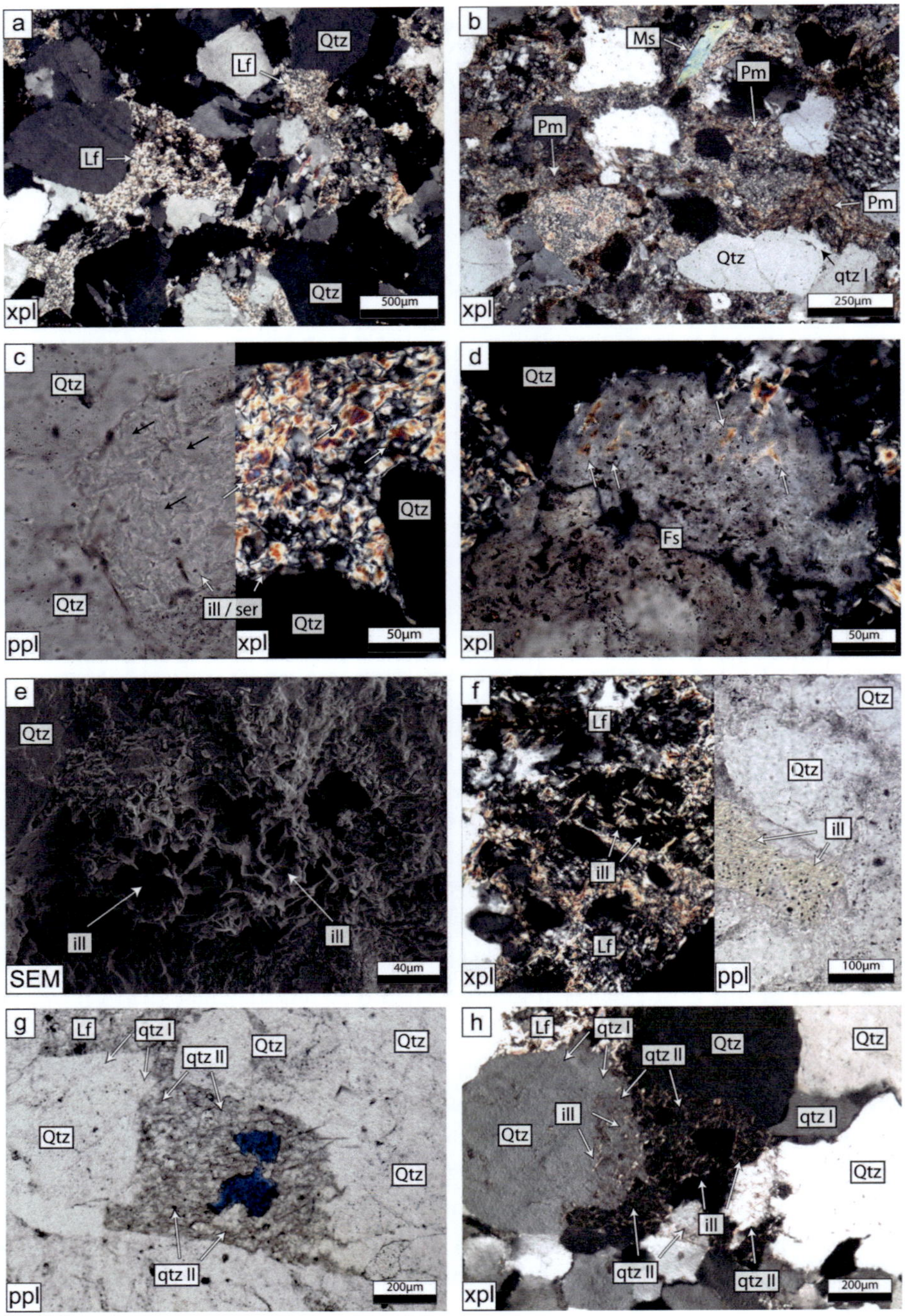

Figure 2.14 (Description on next page)

Figure. 2.14 (Previous page): a) Illitized lithic fragments (Lf), which were squeezed into open pore space. b) Ductile lithic fragments as well as detrital muscovite (Ms) point to pseudomatrix (Pm) formation because of mechanical compaction. Syntaxial quartz (qtz I) overgrows detrital quartz grain (Qtz). c) Crossed-polarized light (xpl) and plane-polarized light (ppl) micrograph of well-ordered booklet shaped aggregates (arrows) of kaolinite transforming into illite (ill). Third-order interference colors suggest a subsequent sericitization (ser) of illite. d) Domains with bright interference colors (arrows) within altered alkali feldspar (Fs) reflect the process of feldspar illitization. e) SEM image of platy pore lining illite (ill) on detrital quartz grain (Qtz). f) A network of fine needles forms meshwork illite (ill) within secondary pore space resulting from the dissolution of detrital components. Note the microporosity highlighted by the yellow epoxy resin. g) & h) Quartz-cemented (qtz II) meshwork illite (ill) formed in secondary pore space. A dark staining (g) of the domain can be recognized, which is the result of an organic/bituminous impregnation of the illite resulting from hydrocarbon charging. The distribution of meshwork illite suggests that it formed in secondary pore space from a dissolved detrital grain (e.g. feldspars). A lithic fragment can be excluded as the quartz cement (qtz II) forms syntaxial overgrowths on adjacent detrital grains with similar extinction in xpl (h). Earlier quartz cement (qtz I) can be distinguished by lesser amounts of fluid inclusions (g). Blue stained porosity (g) demonstrates that the domain is not cemented entirely.

2.5.4.2 Authigenic mineralogy

In all investigated lithologies, authigenic quartz primarily occurs as thin syntaxial overgrowth (qtz I) on detrital quartz grains (Fig. 2.14b). It is difficult to differentiate authigenic quartz from detrital grains as they are often only recognizable by rarely occurring dust rims separating detrital and authigenic quartz or euhedral crystal shapes. Thus, the estimated average authigenic quartz content (9.1 %, Tab. 2.1) of the total rock volume may be an underestimation. In the studied samples, an increased pore-filling quartz cementation is more frequent in coarse-grained samples with lower amounts of pseudomatrix (Tab. 2.1) and near major normal faults (< 35 m; Tab. 2.1). However, the authigenic quartz volume varies and the data are not sufficient to reflect a trend of increasing authigenic quartz volume with decreasing distance to the faults (cf. Tab. 2.1).

The most common authigenic clay mineral in the sandstones is illite. It occurs as well-ordered booklets (Fig. 2.14c), indicating the formation as pseudomorphs because of the alteration of kaolinite (e.g. Bjørlykke and Egeberg, 1993; Worden and Morad, 2003; Bertier et al., 2008). The shape of the illite aggregates suggests that kaolinite occluded secondary pore space, which resulted from alteration of detrital components, such as feldspars. Interference colors and EDX spectra suggest that no pure kaolinite was left and that the aggregates underwent the illitization process. Despite the formation from illitization of clay minerals, illite is also a product of the alteration of lithic fragments and feldspar grains (Fig. 2.14d) and appears as platy, pore-lining flakes (Fig. 2.14e). Meshwork illite (Fig. 2.14f,g,h) is a commonly observed constituent in the samples and is present in most thin sections (Tab. 2.1). The meshwork illite is located in secondary pore space likely resulting from dissolved detrital components. An organic/bituminous staining (cf. Gaupp et al.,

1993) of the meshwork illite domains (Fig. 2.14g) and an intense intergrown quartz cementation (qtz II) can be observed frequently in the samples (Fig. 2.14g,h; Tab. 2.1). The bituminous/organic staining is supported by a measured TOC of 1.5 % for the < 2 µm fraction of a selected sample. A major volume fraction of illite (fibers) was assumed in this extracted size fraction because illite is the most abundant authigenic mineral. Interference colors of all illite variations are unusually high, suggesting a subsequent sericitization (e.g. Fig. 2.14c). The overall illite content based on point-counting data (Tab. 2.1) may be underestimated as the illitized lithic fragments and pseudomatrix (e.g. Fig. 2.14 a,b) are not included in the illite classification. Bulk XRD analyses revealed an average illite/muscovite content of 11 %.

Secondary pores with a characteristic rhombic shape most likely reflect former ankerite (Figs. 2.12b,c; 2.15a). Scanning electron microscopy micrographs illustrate the presence of rhombic ankerite crystals and document the dissolution of the same (Fig. 2.15b). The presence of carbonates is further evidenced by remnants of authigenic ankerite crystals, which are preserved in a few samples (Fig. 2.15c). Therein, carbonate cements appear as pore-space filling and grain-replacing minerals.

Chlorite in the rock matrix (0.5 % on average, Tab. 2.1) occurs as pore-filling aggregates within secondary porosity due to grain dissolution as well as mineral replacement. Chlorite is characterized by a blocky to pseudohexagonal habit within former detrital clasts (Fig. 2.15d). In thin sections, the chlorite is grey to pale green and features anomalous blue interference colors in polarized light (Fig. 2.15d). Scanning electron microscopy uncovered chlorite, which has a typical booklet shape of kaolinite, suggesting a chloritization of kaolinite (Fig. 2.15e). Chlorite flakes occur locally along and within cleaved ankerite cement (Fig. 2.15c) and as small booklet-shaped flakes (Fig. 2.15f). Euhedral chlorite crystals appear within pore-filling quartz cement and are inter-grown with quartz in vein cements with similar characteristic anomalous blue interference colors (Fig. 2.15g,h). Electron microprobe analyses revealed that chlorite in the rock matrix and in the quartz veins have a similar chemical composition with an average Fe/(Fe+Mg) ratio of 0.7 and an average content of tetrahedral aluminum of 2.8 per formula unit (163 individual measurements). As a result of the increased iron content compared to magnesium, the chlorite is classified as chamosite (e.g. Curtis et al., 1985; Okrusch and Matthes, 2014). Based on the tetrahedral aluminum content, the chlorite represents a high-temperature modification (IIb polytype, Curtis et al., 1985) and indicates high-grade diagenetic conditions. Chamosite is thought to precipitate under hydrothermal or anchimetamorphic conditions in iron-rich rocks (e.g. Deer et al., 1992; Spoetl et al., 1994). However, chamosite coats can also form in modern deltaic environments (e.g. Porrenga, 1965; Lemoalle and Dupont, 1973).

Iron (oxide-)hydroxides (2.5 % on average, Tab. 2.1) coat and often occlude the intergranular pore space and are found on lithoclasts. The presence of iron (oxide-)hydroxides is increased in the secondary pore space, probably resulting from iron-bearing carbonate dissolution (Fig. 2.15a). Iron (oxide-)hydroxides are evidenced both by petrographic observations and by bulk XRD analyses.

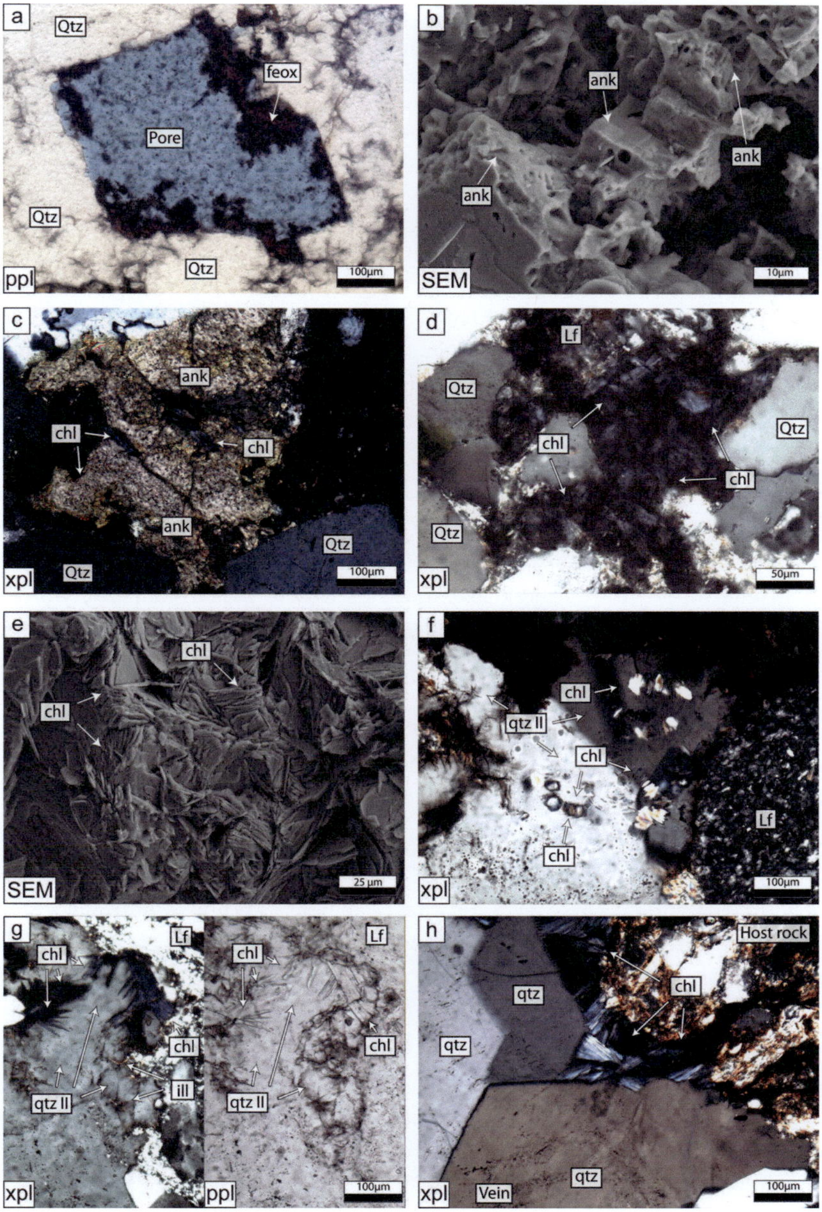

Figure 2.15 (Description on next page)

Figure 2.15 (Previous page): a) Rhombic pore space (Pore) with pore-lining iron (oxide-)hydroxides (feox), which is most likely the result of a dissolved carbonate mineral. b) SEM image of euhedral authigenic ankerite (ank) crystals. Holes in crystal faces indicate the dissolution of ankerite. c) Authigenic ankerite (ank) contains blueish chlorite flakes (chl) along microfractures and cleavage planes. d) Replacing chlorite (chl) within former detrital clasts next to detrital quartz grains (Qtz) and lithic fragments (Lf). The chlorite has blocky to pseudohexagonal shapes. e) SEM image of chlorite (chl) with a typical booklet shape of kaolinite. f) Quartz-cemented pore with chlorite booklets (chl). g) Quartz-cemented (qtz II) euhedral chlorite aggregates and illite fibers (ill) are seen next to lithic fragment (Lf) and detrital quartz (Qtz). h) Entirely sealed quartz-cemented (qtz) vein with syntaxial growth. The quartz is inter-grown with fan-shaped chlorite flakes (chl) at the vein-wall interface.

2.5.4.3 Optical porosity and compaction

Lithic fragments show evidences of alteration and disintegration. Together with mud intraclasts, along with other ductile components, they were subsequently squeezed into open primary pore space. This alludes to mechanical compaction and results in the formation of a pseudomatrix (Fig. 2.14b). The calculated intergranular volume (IGV) allows for the estimation of the contribution of compaction and cementation to the loss of primary porosity (e.g. Ehrenberg, 1995; Paxton et al., 2002). The formation of a pseudomatrix and an average IGV of 12.3 % (Tab. 2.1) supplies support for a strong impact of mechanical compaction. This is further corroborated by the COPL of 78.5 % on average (Tab. 2.1). Chemical compaction is evidenced by sutured or concave-convex grain contacts (Fig. 2.14a,h). Along grain-grain interfaces with micas, quartz minerals feature planar contacts. Moreover, all calculated IGV values are significantly below 26 % (Tab. 2.1), which evidences additionally the involvement of chemical besides mechanical compaction (e.g. Ehrenberg, 1995).

All of the observed macroporosity is secondary and of both intra- and intergranular origin, a consequence of the dissolution of authigenic and detrital components. The characteristic rhombic pores (Fig. 2.15a) manifest the contribution of carbonate dissolution besides the dissolution of feldspars and alumosilicates. In pseudomatrix-bearing samples, deformed ductile lithic fragments occlude primary pore space (Fig. 2.14a,b). Authigenic quartz cements occlude the remaining primary pore space as well as primary pore space in samples without large amounts of pseudomatrix. In most of the thin sections, observable secondary micropores of illite meshes are substantially cemented with authigenic quartz (qtz II, Fig. 2.14g,h; Tab. 2.1). The influence of authigenic minerals on porosity evolution is reflected in the calculated cementational porosity loss (CEPL) of 21.5 % on average (Tab. 2.1). The optical porosity (4.6 % on average, Tab. 2.1) derived by point-count analysis is constantly lower than porosity derived by helium pycnometry (8.5 % on average, Tab. 2.1). This suggests a significant contribution of microporosity (4.2 % on average, Tab. 2.1).

Table 2.1: Mineralogical compositions of 51 selected samples of the Upper Westphalian C and D derived by point-counting based on 300 counts per thin section. IGV, COPL and CEPL were calculated. Helium porosity measurements resulted in higher values compared to optical porosity and the contribution of microporosity (µ-poro) was calculated by subtracting secondary macroporosity.

	Sample ID	Point counting results [%]																								Petrophysics			Grainsize [mm]			
		Detrital components										Authigenic components											IGV									
		Qm	Qp	Fsp	Mica	Sst RF	Fsp Sst RF	MRF	Cl RF	RF indet	PM	Qa	Qa, Ill pf	Ill rp Kaol	Ill pf	Ill rp FS	Ill ind	Chl	Clay ind	Feox	Fe Carb	Poro	IGV	COPL	CEPL	He Poro [%]	Perm [mD]	µ-poro [%]	Average	Max	Cat	Sorting
Distal to W-E striking faults (> 100 m)	SS02	43.0	6.7	1.7	0.3	0.0	0.0	0.0	1.3	5.7	1.0	7.0	0.7	4.0	0.3	0.0	2.7	0.0	0.0	7.0	0.0	18.7	14.3	74.9	25.1	15.7	0.00400	-	0.38	0.69	mS	mod
	SS03	47.0	12.3	0.3	0.0	2.0	1.0	0.7	0.7	5.3	3.0	10.7	7.0	3.3	0.7	0.0	3.7	0.0	0.0	1.7	0.0	0.7	13.0	77.6	22.4	7.2	0.00022	6.5	0.47	0.94	mS	mod
	SS05	59.3	15.0	2.7	0.0	4.7	1.3	0.0	6.3	0.0	2.3	0.7	0.7	0.3	0.0	0.0	0.0	0.0	0.7	2.3	0.0	3.7	3.0	95.4	4.6	8.3	0.00021	4.7	0.45	1.11	mS	mod
	SS09	53.3	10.7	0.0	0.0	1.0	1.3	0.0	1.7	5.3	3.3	8.3	2.0	0.0	0.3	0.0	2.7	0.0	0.0	3.3	0.0	6.7	12.0	79.5	20.5	10.3	0.00081	3.6	0.43	1.12	mS	mod
	SS13	47.3	14.3	0.7	0.7	3.0	1.7	1.0	1.3	5.7	5.3	6.3	2.3	0.7	0.7	0.3	2.3	0.0	0.0	2.3	0.0	4.0	9.3	84.6	15.4	5.9	0.00093	1.9	0.46	0.96	mS	mod
	SS16	46.3	3.7	2.3	1.3	0.0	3.7	0.0	1.3	3.3	33.7	1.3	0.0	0.0	0.0	1.0	2.0	0.0	0.0	0.0	0.0	0.0	1.3	98.0	2.0	0.2	0.00002	0.2	0.34	0.56	mS	mod
	SS18	64.0	4.3	0.7	0.0	0.0	0.0	0.0	1.3	6.0	8.7	5.7	1.3	1.3	2.7	0.0	1.0	0.0	0.0	2.0	0.0	1.0	10.3	82.7	17.3	8.6	0.00036	7.6	0.46	0.84	mS	mod
	SS22a	41.0	12.7	0.3	0.3	0.7	0.7	1.0	3.3	7.7	4.7	10.0	5.0	1.3	1.0	0.0	4.3	0.0	0.7	2.3	0.0	3.0	13.3	76.9	23.1	6.6	0.00011	3.6	0.46	0.94	mS	mod
	SS23	47.0	15.3	0.3	0.0	2.3	3.0	2.7	1.0	9.3	2.0	8.7	1.7	0.7	0.0	0.0	1.7	0.0	0.0	2.7	0.0	1.7	11.3	80.8	19.2	6.6	0.00012	4.9	0.56	1.30	cS	mod well
	SS25	67.3	10.7	1.7	0.7	1.7	0.3	1.7	4.3	0.0	1.7	1.0	1.0	0.7	0.0	0.0	1.0	0.0	0.3	5.0	0.0	1.0	6.0	90.4	9.6	7.2		6.2	0.46	0.94	mS	mod
	SS26	55.0	4.7	0.3	0.0	0.0	0.0	0.0	0.3	7.0	0.7	6.0	2.7	0.3	9.3	0.0	1.7	0.0	0.0	7.3	0.0	4.7	22.7	56.0	44.0	11.6	0.00003	6.9	0.34	0.60	mS	mod
	SS27	51.7	7.3	1.3	0.0	0.3	0.7	0.3	1.0	5.3	8.3	8.3	1.7	1.0	0.3	0.0	2.7	0.0	1.0	0.0	0.0	8.7	8.7	85.8	14.2	11.5	0.00039	2.8	0.39	0.94	mS	mod
	SS28	46.7	6.3	1.0	0.3	1.3	2.0	1.7	0.7	7.3	4.3	6.3	0.0	1.7	0.0	1.0	2.7	0.0	0.7	8.3	0.0	7.7	14.7	74.2	25.8	14.7	0.02900	7.1	0.47	2.34	mS	mod well
	SS33	42.3	7.0	0.7	0.3	1.7	3.0	1.0	1.0	4.3	17.7	3.3	0.0	2.3	0.7	0.0	4.3	0.0	1.7	2.7	0.0	6.0	6.7	89.3	10.7	9.6		3.6	0.30	1.20	mS	mod
	MK03	50.7	5.3	0.3	0.7	0.3	2.0	1.7	1.7	8.7	3.0	7.0	5.0	2.3	1.0	0.0	2.3	0.0	1.0	3.3	0.0	3.7	11.3	80.8	19.2	4.6	0.00006	0.9	0.29	0.54	mS	mod
	MK04a	57.7	10.0	2.7	1.0	1.3	2.0	1.0	6.7	0.0	4.0	1.3	0.7	0.0	0.0	0.0	0.7	0.0	0.0	8.3	0.0	2.7	9.7	83.9	16.1	8.7	0.00045	6.1	0.33	0.55	mS	mod
	MK05	42.3	5.7	0.3	0.0	0.0	0.0	0.0	2.3	4.7	2.7	6.7	0.0	1.7	0.7	0.0	3.3	0.0	0.0	11.3	0.0	18.3	18.7	65.6	34.4	19.9		1.5	0.32	0.72	mS	mod
	MK06	44.7	10.7	1.0	0.0	3.0	1.7	0.7	0.7	4.3	2.0	12.3	3.0	2.0	1.7	1.0	3.3	0.0	0.3	3.7	0.0	4.0	17.7	67.8	32.2	7.0	0.00053	3.0	0.34	0.54	mS	mod
	MK08	73.0	5.0	1.0	0.7	3.0	0.7	1.3	5.7	0.7	3.0	0.0	2.3	0.0	0.0	0.0	0.3	0.0	0.3	2.3	0.0	0.7	2.3	96.4	3.6	6.2	0.00036	5.5	0.42	0.82	mS	mod
	MK09	53.3	6.7	0.7	0.7	0.7	1.7	1.7	0.7	3.7	5.3	9.7	4.7	1.3	1.0	0.0	3.7	0.0	0.3	3.7	0.0	0.7	14.3	74.9	25.1	6.5	0.00028	5.8	0.32	0.62	mS	mod
	MK11	52.0	4.7	0.7	0.0	1.7	1.3	1.3	1.3	4.3	8.0	6.0	5.0	2.3	1.0	0.0	4.0	0.0	0.7	2.0	0.0	3.7	9.0	85.2	14.8	6.0	0.00046	2.3	0.36	0.71	mS	mod
	MK12	47.0	10.0	0.3	0.0	2.7	2.0	0.7	1.0	5.7	2.7	11.0	4.0	3.7	0.3	0.0	4.7	0.0	0.3	1.7	0.0	2.3	13.0	77.6	22.4	5.0	0.00038	2.7	0.39	0.88	mS	mod
	MK13	56.0	8.3	0.3	0.0	0.3	1.0	1.3	1.0	3.3	3.3	8.7	1.7	1.3	0.3	0.0	7.3	0.0	0.3	2.7	0.0	2.7	11.7	80.2	19.8	6.0	0.00015	3.3	0.31	0.66	mS	mod
	MK14	55.0	6.0	0.0	0.0	0.7	0.0	0.0	3.0	10.3	2.7	10.3	5.0	2.3	0.0	0.0	3.3	0.0	0.0	1.3	0.0	0.0	11.7	80.2	19.8	8.1	0.00029	8.1	0.39	0.82	mS	mod
	MK16	52.7	1.3	1.3	0.3	0.0	0.0	0.0	1.7	10.0	7.3	15.0	4.0	2.7	0.0	0.0	2.0	0.0	0.0	0.3	0.7	0.7	16.0	71.4	28.6	5.2		4.5	0.41	0.74	mS	mod
	MK17	55.0	8.3	0.7	0.7	1.0	1.0	1.0	1.3	3.3	4.0	6.0	5.7	2.0	1.0	0.0	4.3	0.0	0.7	1.7	0.0	2.3	8.7	85.8	14.2	5.7	0.00019	3.4	0.36	0.84	mS	mod
	MK24	44.3	10.7	0.0	0.3	0.3	0.0	0.0	4.0	5.7	1.3	16.7	7.0	3.7	0.3	0.0	1.3	0.0	0.0	3.7	0.0	0.7	20.7	60.9	39.1	5.0	0.00033	4.3	0.47	0.86	mS	mod
	Average	*51.7*	*8.3*	*0.9*	*0.3*	*1.2*	*1.2*	*0.8*	*2.1*	*5.1*	*5.4*	*7.2*	*2.7*	*1.6*	*0.9*	*0.1*	*2.7*	*0.0*	*0.3*	*3.4*	*0.0*	*4.1*	*11.5*	*79.9*	*20.1*	*8.1*	*0.0003*	*4.3*	*0.40*	*0.88*	-	-

Table 2.1 (Continued)

	Sample ID	Point counting results [%]																								Petrophysics			Grainsize [mm]			
		Detrital components										Authigenic components											IGV									
		Qm	Qp	Fsp	Mica	Sst RF	Fsp Sst RF	MRF	Cl RF	RF indet	PM	Qa	Qa, Ill pf	Ill rp Kaol	Ill pf	Ill rp FS	Ill ind	Chl	Clay ind	Feox	Fe Carb	Poro	IGV	COPL	CEPL	He Poro [%]	Perm [mD]	µ-poro [%]	Average	Max	Cat	Sorting
Proximal to W-E striking faults (< 50 m)	PB001	41.7	1.0	0.7	1.0	0.0	0.0	0.0	5.3	10.0	13.7	11.0	0.0	0.0	0.0	0.0	2.7	7.7	0.0	0.0	0.0	5.3	11.0	81.5	18.5	14.5	0.23163	9.2	0.14	0.23	fS	mod
	PB008 (11 m)	51.0	6.7	1.3	0.0	0.0	0.0	0.0	5.7	7.0	4.7	10.3	5.3	1.3	0.0	0.0	3.0	1.7	2.0	0.0	0.0	0.0	10.3	82.7	17.3	1.8	0.00012	1.8	0.36	0.71	mS	mod
	PB034 (14 m)	46.0	6.7	0.7	0.0	0.7	0.0	0.7	0.7	12.0	2.7	14.0	3.0	2.0	0.0	0.0	3.3	3.3	0.0	1.7	0.0	2.7	15.7	72.1	27.9	8.7	0.00019	6.0	0.44	0.90	mS	mod
	PB009 (15 m)	54.0	4.3	0.3	0.0	0.0	0.0	0.0	1.7	6.0	4.7	11.0	8.3	2.0	0.0	0.0	3.3	2.0	0.0	0.0	0.0	2.3	11.0	81.5	18.5	6.0	-	3.7	0.55	1.02	cS	mod
	PB019 (15.5 m)	49.3	9.7	0.7	0.0	0.0	0.0	0.0	0.7	8.3	0.7	14.3	8.0	2.0	0.0	0.0	2.3	3.0	0.0	1.0	0.0	0.0	15.3	72.8	27.2	4.3	0.00065	4.3	0.47	0.97	mS	mod
	PB018 (16 m)	56.7	8.3	0.3	0.3	0.0	0.0	0.0	1.7	4.7	0.0	12.3	6.0	2.0	0.0	0.0	1.7	3.7	0.0	0.7	0.0	1.7	13.0	77.6	22.4	NA	-	0.0	0.50	1.11	cS	mod
	PB006 (18 m)	45.3	10.0	0.3	0.0	0.0	0.0	0.0	6.0	5.3	1.0	13.0	12.0	1.0	0.3	0.0	1.3	3.3	0.3	0.7	0.0	0.0	14.0	75.6	24.4	2.8	-	2.8	0.57	1.52	cS	mod well
	PB004 (24 m)	56.3	7.0	0.7	0.0	0.3	0.0	0.0	1.7	4.3	1.7	13.7	7.7	1.0	0.0	0.0	1.3	1.3	0.7	0.0	0.0	2.3	13.7	76.3	23.7	3.3	-	1.0	0.62	1.77	cS	mod well
	PB014 (27 m)	44.3	4.0	0.7	1.0	0.0	0.0	0.0	4.0	6.0	18.0	10.0	1.3	3.7	0.0	0.0	2.7	0.0	0.3	3.0	0.0	1.0	13.0	77.6	22.4	0.5	0.00009	0.0	0.27	0.51	mS	mod
	PB068	48.0	6.7	2.0	0.3	0.7	0.0	0.3	2.3	8.0	3.0	6.7	0.7	2.0	0.0	0.0	3.7	0.0	0.0	5.3	0.0	10.3	12.0	79.5	20.5	16.5	-	6.1	0.40	1.10	mS	mod
	PB069	49.3	6.0	0.7	0.0	0.3	0.7	0.7	2.3	3.7	5.3	10.0	7.7	0.3	1.0	0.0	5.3	0.7	2.0	0.0	0.0	4.0	11.0	81.5	18.5	7.2	0.00028	3.2	0.39	1.99	mS	mod
	PB073	54.3	9.7	0.7	0.7	1.7	1.0	0.7	0.7	2.7	8.3	9.3	4.7	0.7	0.0	0.0	4.0	0.3	0.0	0.0	0.0	0.7	9.3	84.6	15.4	4.2	0.00005	3.5	0.31	0.58	mS	mod
	PB075	50.0	5.7	1.0	0.7	0.7	2.0	1.0	2.3	4.0	11.3	11.3	3.0	0.3	0.7	0.0	2.3	0.0	0.7	0.7	0.0	2.3	12.7	78.2	21.8	5.8	0.00004	3.4	0.30	0.84	mS	mod
	PB080	44.3	9.3	0.7	0.3	1.0	2.0	1.7	2.3	4.7	1.7	12.3	5.3	2.3	0.3	0.0	5.7	0.0	0.0	4.3	0.0	1.7	17.0	69.3	30.7	13.1	0.00004	11.4	0.36	0.66	mS	mod
	R20.0	36.3	7.3	1.0	0.3	0.7	1.3	0.3	1.0	5.7	6.0	12.0	6.7	1.7	1.7	0.0	5.3	0.0	0.7	0.0	0.0	12.0	13.7	76.3	23.7	15.1	0.65966	3.1	0.34	0.65	mS	mod
	Average	*48.5*	*6.8*	*0.8*	*0.3*	*0.4*	*0.5*	*0.4*	*2.6*	*6.2*	*5.5*	*11.4*	*5.3*	*1.5*	*0.3*	*0.0*	*3.2*	*1.8*	*0.4*	*1.2*	*0.0*	*3.1*	*12.8*	*77.8*	*22.2*	*7.4*	*0.0002*	*4.0*	*0.40*	-	-	-
Fault (damage) zones (< 5 m)	PB011 (0 m)	52.0	6.3	0.7	0.0	0.0	0.0	0.0	2.3	3.7	1.3	16.7	1.7	3.0	0.0	0.0	2.3	1.0	0.0	5.7	0.0	3.3	22.3	56.9	43.1	7.3	-	4.0	0.49	1.19	mS	mod
	PB013 (0 m)	50.7	5.3	0.7	0.3	0.0	0.0	0.0	3.7	9.3	3.3	16.7	3.0	2.0	0.0	0.0	2.0	0.0	1.0	0.3	0.0	1.7	17.0	69.3	30.7	4.0	-	2.3	0.42	0.77	mS	mod
	PB098	52.3	4.7	0.7	0.0	0.7	2.0	0.0	3.7	3.0	1.7	11.7	0.0	0.0	0.0	0.0	0.3	0.0	0.0	0.0	0.0	19.3	11.7	80.2	19.8	NA	-	0.0	0.44	0.81	mS	mod
	PB099	43.7	4.3	0.0	0.0	0.7	2.0	0.0	2.7	5.3	3.0	11.7	0.0	0.0	0.0	0.0	0.0	0.0	0.0	0.0	0.0	26.7	11.7	80.2	19.8	26.3	104.70	0.0	0.44	0.84	mS	mod
	PB100	36.3	5.7	0.7	0.3	0.7	2.7	2.0	4.0	4.0	7.0	10.7	3.7	1.7	3.7	0.7	5.0	0.0	1.7	0.0	0.0	9.7	14.3	74.9	25.1	14.4	0.15000	4.7	0.34	0.67	mS	mod
	L2.0	45.3	9.0	0.3	0.0	0.3	0.0	0.0	0.7	13.0	0.0	14.0	6.7	4.3	0.3	0.0	4.0	0.0	0.0	0.0	0.0	2.0	14.3	74.9	25.1	8.8	-	6.8	0.37	0.80	mS	mod
	R0.75	50.3	10.3	1.7	4.3	3.7	0.7	1.0	7.0	2.0	4.7	1.0	4.3	0.0	0.0	0.0	0.0	0.0	2.3	4.7	0.0	2.0	5.7	91.0	9.0	11.9	-	9.9	0.29	0.75	mS	mod
	R1.0	47.0	10.7	0.0	0.0	0.0	0.0	0.0	1.7	9.3	6.7	11.7	7.0	1.3	0.0	0.0	3.3	0.0	0.0	1.0	0.0	0.3	12.7	78.2	21.8	1.0	-	0.7	0.36	1.07	mS	mod
	R2.0	46.3	6.0	0.3	0.0	0.0	0.0	0.0	1.7	9.0	3.7	5.7	0.3	1.7	3.0	0.0	4.0	0.0	0.0	5.0	0.0	13.3	13.7	76.3	23.7	23.6	24.93000	10.3	0.33	0.75	mS	mod
	Average	*47.1*	*6.9*	*0.6*	*0.6*	*0.7*	*0.8*	*0.3*	*3.0*	*6.5*	*3.5*	*11.1*	*3.0*	*1.6*	*0.8*	*0.1*	*2.3*	*0.1*	*0.6*	*1.9*	*0.0*	*8.7*	*13.7*	*75.8*	*24.2*	*12.2*	*24.9300*	*4.3*	*0.39*	*0.85*	-	-
Overall average		**49.9**	**7.6**	**0.8**	**0.4**	**0.9**	**0.9**	**0.6**	**2.4**	**5.6**	**5.1**	**9.1**	**3.5**	**1.6**	**0.7**	**0.1**	**2.8**	**0.5**	**0.4**	**2.5**	**0.0**	**4.6**	**12.3**	**78.5**	**21.5**	**8.5**	**0.0003**	**4.2**	**0.40**	**0.90**	-	-

Explanation **Table 2.1** (On next page)

Explanation **Tab. 2.1** (Previous pages): Qm = quartz, monocrystalline; Qp = quartz, polycrystalline; Fsp = feldspar; Sst RF = sandstone rock fragment; Fsp Sst RF = feldspathic sandstone rock fragment; MRF = metamorphic rock fragment; Cl RF = shale/claystone rock fragment; RF ind = rock fragment, indeterminate; PM = pseudomatrix; Qa = authigenic quartz; Qa, ill pf = quartz-cemented pore-filling illite (meshwork); ill rp kaol = illite replacing kaolinite; ill pf = pore-filling illite (meshwork); ill rp Fs = illite replacing feldspar; ill ind = illite, indeterminate; Chl = chlorite; clay ind = clay, indeterminate; Feox = iron (oxide-)hydroxides; Fe carb = iron carbonate (ankerite); fS = fine-grained-sandstone; mS = medium-grained sandstone; cS = coarse-grained sandstone; mod = moderately sorted; mod well = moderately well sorted. Sample IDs with distance measures in parenthesis belong to the same sampling profile and the measures indicate the distance to the next fault. Although observing overall increased quartz cementation in fault areas (< 50 m to fault), no distinct trend of increasing quartz content with decreasing distance to the fault can be derived.

2.5.4.4 Correlation of petrography and petrophysics

The primary composition and diagenetic overprint during the burial history control the reservoir quality of sandstones. The composition is a direct result of factors like the depositional system, -environment and the provenance area. Burial cements, such as quartz and illite, have a major impact on reservoir quality evolution (e.g. Dutton et al., 1991; Bjørlykke and Egeberg, 1993; Worden and Morad, 2003). In order to assess the reservoir quality controlling factors, the quantitative volumetric data from the point-counting analysis and petrographic grain-size analysis were correlated with the porosity and permeability measurements (Tab. 2.1, Fig. 2.16). This includes the main authigenic minerals, detrital components as well as grain-size analyses and IGV calculations (Fig. 2.16). The results show that no meaningful correlations can be derived from the cross-plots. Porosity and permeability plots have values for R^2 significantly below 0.1. The regression for permeability against average grain size has a slightly better R^2 value of 0.136 (Fig. 2.16). Iron (oxide-)hydroxides have a positive correlation with porosity at an R^2 of 0.455 (Fig. 2.16). However, the preliminary occurrence of iron (oxide-)hydroxides in secondary pore space likely stemmed from dissolved carbonates (cf. Fig. 2.15a). Hence, secondary porosity is a necessary criterion for the presence of iron (oxide-)hydroxides and consequently does not reflect a reservoir-controlling relationship.

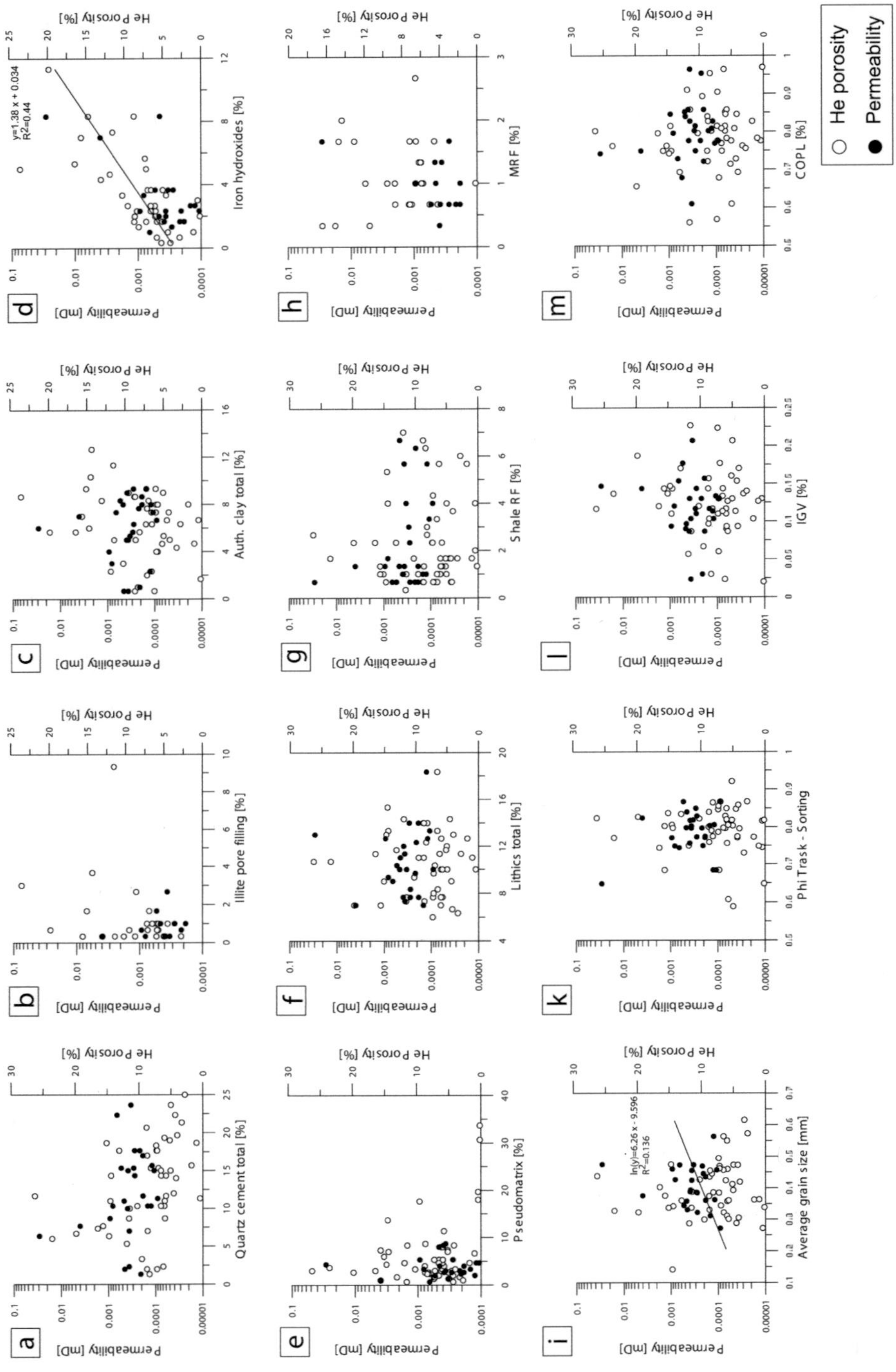

Figure 2.16 (Description on next page)

Figure 2.16 (Previous page): Correlation plots based on point-count data (Tab. 2.1) of helium porosity and permeability against a) the total amount of authigenic quartz, b) pore-filling illite (meshwork), c) the total amount of authigenic clay, d) iron (oxide-)hydroxides, e) pseudomatrix, f) total amount of lithic fragments, g) shale rock fragments, h) metamorphic rock fragments, i) average grain size, k) sorting, l) IGV and m) COPL. No distinct correlation trends can be established except for the content of iron (oxide-)hydroxides against helium porosity (d). Regression functions with R^2 values below 0.1 are not reported.

2.6 Discussion

2.6.1 Sedimentology

The vertical profile comprises four stratigraphic fourth-order cycles (sequence or Milankovitch band cycles, Süss et al., 2000), identified by large erosive structures between fine- and coarse-grained rocks (Fig. 2.3). The exposed 135 m strata in the outcrop does not permit the delineation of a lower grade third-order cyclicity (cf. Süss et al., 2000), while fourth-order cycles in the study area range from 20 to 50 m.

Coarse-grained channel lags at the base of individual fining-upwards cycles (e.g. Fig. 2.5b) are generally interpreted as high- to medium-energy downstream accretion dominated systems (cf. Jones and Glover, 2005), and this is also indicated by high-angle trough cross-bedding and mid-channel bar forms (Fig. 2.5a). Low-energy lateral accretion dominated channels composed of medium- to fine-grained sediments at the upper part of each cycle (Fig. 2.3) are characterized by low-angle cross-bedding (e.g. Fig. 2.5d). Therefore, the sandstones originated predominantly from braided river deposits (David, 1990). One coarsening upward sequence is located between the „Dreibänke“ and „Mittel“ coal seams (q.v. David, 1987; Jones and Glover, 2005). This approximately 10 m thick sequence can be interpreted as crevasse splay (e.g. David, 1987) or lacustrine delta (e.g. Jones and Glover, 2005). The fining-upwards and coarsening-upwards cycles are well-reflected by gamma ray data (Fig. 2.3) and match with subsurface data from the tight gas fields (e.g. Wimmers and Koehrer, 2014).

2.6.2 Diagenetic evolution/paragenetic sequence

Observations from petrographic analysis were interpreted and summarized in a paragenetic sequence (Fig. 2.17) in order to highlight the most important diagenetic processes in relation to the burial history of the study area.

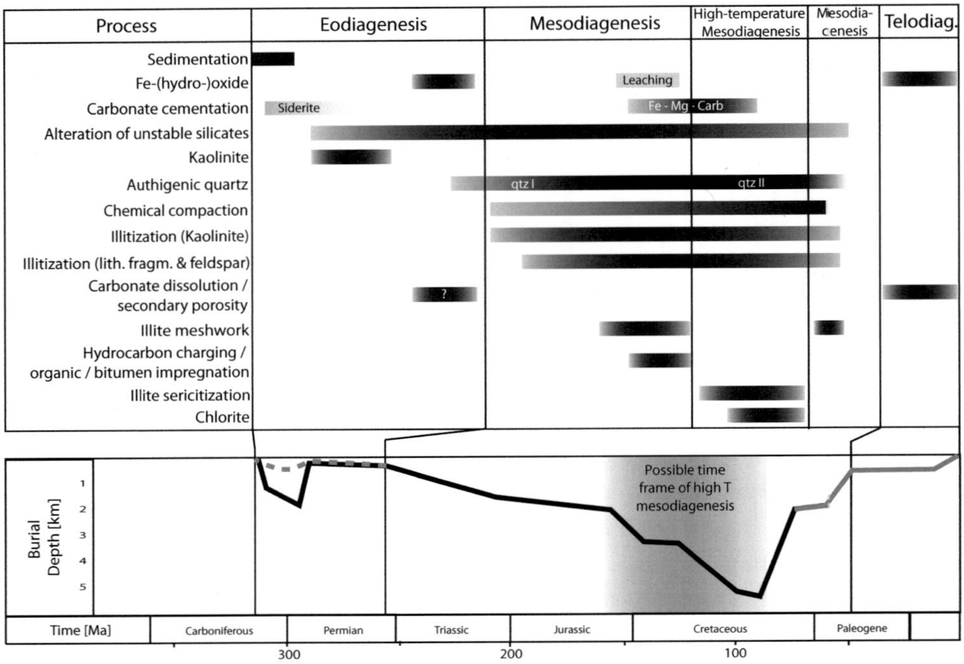

Figure 2.17: Paragenetic sequence based on petrographic analysis underscoring the most significant diagenetic processes with respect to the different diagenetic regimes as described in the text. The paragenetic sequence is associated with the basin evolution and coupled to modeled burial curves (black curve represents burial history of the Lower Saxony Basin based on Bruns et al., 2013; grey curve represents burial history in uplift phase of the Piesberg based on Senglaub et al., 2005). The burial graph was modified for the studied outcrop (dashed grey line) as no petrographic indications were identified for a first deep burial reaching a mesogenetic realm and subsequent uplift to telogenetic conditions in the Early Permian. The burial history of the Lower Saxony Basin after Bruns et al. (2013) is considered also representative for Upper Carboniferous subsurface reservoirs in the region, except for the amount of uplift and exposure to the surface. Upper Carboniferous reservoirs are presently at a depth between approximately two to five km in the Lower Saxony Basin.

2.6.2.1 Eodiagenesis

Eodiagenesis is controlled by the climate, the depositional environment and the primary rock composition (Worden and Morad, 2003). The main eogenetic processes include mechanical compaction and alteration of unstable silicates as well as kaolinite authigenesis.

The formation of kaolinite (e.g. Fig. 2.14c) as a result of the alteration of feldspars and other detrital alumosilicates can be attributed to the humid climate (Süss et al., 2000; Roscher and Schneider, 2006; Drozdzewski et al., 2008) as well as to low-pH groundwater and constant flushing with meteoric waters. The kaolinitization led to the development of secondary microporosity (Füchtbauer, 1974; Worden and Morad, 2003) and may have been a first source of silica for the precipitation of minor amounts of authigenic quartz in late eodiagenesis. However, due to temperatures below 70 – 80°C in the eogenetic regime, the precipitation rate of quartz is very low (Bjørlykke and Egeberg, 1993) and thus the contribution of eogenetic quartz was less significant.

Lithic fragments were altered and disintegrated. Together with mud intraclasts and similar ductile components, they were squeezed into primary open pore space because of mechanical compaction (e.g. Fig. 2.14a,b). Similar observations were made for the Southern North Sea Basin (Leeder and Hardman, 1990) and the Campine Basin in Belgium (Bertier et al., 2008). This resulted in the development of a pseudomatrix, causing an early loss of primary porosity (Worden and Morad, 2003).

2.6.2.2 Mesodiagenesis

The mesogenetic realm began during deeper burial with temperatures > 70°C (e.g. Morad et al., 2000). The most important processes observed in the Upper Carboniferous sandstones of the Piesberg quarry include chemical compaction, authigenic quartz precipitation, illitization and ankerite authigenesis.

Permian uplift is frequently reported for the region (e.g. Senglaub et al., 2006; Bruns et al., 2013). The estimated minimum thickness of the Westphalian D in the area is 300 m (e.g. Drozdzewski et al., 2009). The Westphalian D of the studied quarry has a thickness of 90 m (Fig. 2.3a). As it is unconformably overlain by Zechstein carbonates (Voigt, 1960), a minimum erosion of 200 m can be assumed. No petrographic evidence were found, which indicate a deeper burial reaching a mesogenetic temperature regime prior to Permian uplift. However, this does not imply the absence of such a first deeper burial phase.

Authigenic quartz precipitation, beginning in late eodiagenesis, is favored at temperatures above 70 – 80°C (Bjørlykke and Egeberg, 1993) forming quartz overgrowth (qtz I). Silica is released from abundant sutured and concave-convex quartz-quartz contacts as well as straight quartz-mica contacts (e.g. Fig. 2.14a,h). Another source for silica is clay mineral authigenesis, such as the illitization of kaolinite (Fig. 2.14c) and feldspar-bearing clasts (Fig. 2.14d) (Haszeldine et al., 1984; Worden and Morad, 2003).

The generation of illite meshes is a consequence of the illitization of detrital grains, like feldspars and lithic fragments, indicated by angular outlines (e.g. Fig. 2.14g,h) and feldspar fragments. A significant illitization of feldspars takes place at temperatures higher than 130°C (Worden and Morad, 2003) or even 150°C (Liewig and Clauer, 2000). Such temperatures may be correlated with the burial curve (Fig. 2.17) for the estimation of the timing of meshwork illite generation and associated bituminous impregnation (Fig. 2.14g), which formed since Late Jurassic times. The bituminous impregnation may be correlated with hydrocarbon charging from the coal measures generating gas at 1 to 3 vitrinite reflectance values (Gautier, 2003, p.16). Hydrocarbon charging from the mid-Triassic until inversion times during the Late Cretaceous (Adriasola-Muñoz et al., 2007) is akin to the Rotliegend reservoir charges in the North Sea (Sole Pit basin 159 ± 16 Ma, Robinson et al. 1993; Broad Fourteens basin 140 Ma, Gautier 2003).

Bitumen-stained illite meshwork (Fig. 2.14g) is generally sealed by a later phase of intense quartz cementation (qtz II, Fig. 2.14h), which is prevalent in coarse-grained sandstones (Tab. 2.1) and commences following hydrocarbon charging (Fig. 2.17). The decomposition of minerals, such as feldspar, continued during ongoing subsidence forming uncemented illite meshwork (Fig. 2.14f), which are dominant in medium- to fine-grained sandstones as well as in small pores of coarse-grained sandstones.

Ankerite is a common mesogenetic carbonate cement within secondary pores and can originate from the transfer of iron and magnesium from dissolving unstable silicates, such as clay minerals or feldspars (Boles, 1978; Kantorowicz, 1985). However, the occurrence of ankerite in sedimentary rocks is also frequently a result of hydrothermal alteration as well as low-temperature metasomatism (Deer et al., 1992). Based on the euhedral rhomb geometry also replacing quartz (Fig. 2.15c), we conclude there was ankerite formation after quartz I overgrowth.

The existence of the late intense quartz cementation (qtz II), the association with chlorite and sericitized illite as well as the euhedral shape of ankerite all implies high temperatures during mesodiagenesis (Fig. 2.17). This can be related to deep burial or a high-temperature event. Chlorite elemental composition establishes a high-temperature modification of chamosite, outlining a high-grade diagenetic regime. Reported homogenization temperatures of quartz vein fluid inclusions trapped along with saline fluids support the petrographic observations of a high-temperature exposure of the studied rocks of at least 250°C (q.v. Chapter 3).

Chlorite arranged within cleaved ankerite (Fig. 2.15c) infers ankerite prior to chlorite formation. Such high temperatures of the studied outcrop are also manifested in the high maturity of organic matter with a high coalification rank of meta-anthracite (e.g. Bässler and Hoyer, 1971; Teichmüller et al., 1984; Betzer et al., 2003). The high temperatures in the study area were related to a plutonic intrusion (e.g. Bartenstein et al., 1971; Betz et al., 1987), and recently to a deep burial of more than 7 km and inversion with subsequent uplift (e.g. Petmecky et al., 1999; Senglaub et al., 2006; Bruns et al., 2013). The high temperatures may also be interpreted as a product of the circulation of hydrothermal fluids (q.v. Chapter 3), evidenced by the presence of chamosite.

2.6.2.3 Telodiagenesis

Telodiagenesis takes place in uplifted rocks, which are controlled by the surface environment under the influence of percolating meteoric surface waters (Schmidt and McDonald, 1979). Rocks investigated at the Piesberg have undergone telogenetic alterations since at least the Early Tertiary (Fig. 2.17).

The missing Rotliegend and the unconformably overlying Zechstein (Voigt, 1960) point to an Early Permian exposure of Upper Carboniferous rocks. Thus, rocks should have been affected by an overprint resulting from an arid Permian environment. Quartz-overgrown iron oxide-stained rims of detrital quartz grains (e.g. Turner et al., 1995) were not observed in the quarry. Their formation from ferric mineral oxidation would point to a first telogenetic phase (Sedat, 1992). Iron oxide rims may have been subsequently leached by hydrocarbons during later mesodiagenesis, as described for Rotliegend sediments by Schöner and Gaupp (2008).

After the Late Cretaceous basin inversion phase and subsequent uplift of the Piesberg to near-surface depths, telogenetic processes altered the strata. According to Senglaub et al. (2005), the study area was at < 60°C and hence in the telogenetic zone since Paleocene/Eocene times. The most important process with respect to porosity evolution is carbonate dissolution besides the dissolution of unstable alumosilicates. This is manifested in the characteristic rhomboidal pores and associated iron (oxide-)hydroxide precipitation, which is a result of iron-rich carbonate alteration (Fig. 2.15). The correlation between iron (oxide-)hydroxides and porosity (Fig. 2.16i) implies that secondary porosity is a necessary criterion for the occurrence of iron (oxide-)hydroxides. Therefore, it is likely that they are a proxy for former authigenic carbonates in associated pores and an evidence of the same. Euhedral carbonates formed pre-peak subsidence (Fig. 2.17) and dissolved grain pores still have a rhombic shape. Hence, and because the pores lack other mesogenetic precipitates, such as a second authigenic quartz phase, ankerite dissolved post-peak subsidence.

2.6.3 Reservoir quality evolution

Reservoir quality is controlled by a combination of the depositional system, diagenetic overprint and structural history. Sedimentological features as well as the measured porosities (7 % on average) match the subsurface data pertaining to the tight gas reservoirs of the region (Fig. 2.8). However, the subsurface data show permeabilities of approximately one to two orders of magnitude higher than the outcrop data (Fig. 2.8).

The correlation between petrographic data with that of porosity and permeability (Fig. 2.16) yields no distinct reservoir controlling trends. This is related to the mesogenetic high-temperature event, resulting in intense quartz cementation and illitization with a reduction of porosity and permeability (e.g. Bjørlykke and Egeberg, 1993; Morad et al., 2010). While rocks of the studied outcrop were likely exposed to temperatures exceeding 250°C (q.v. Chapter 3), the subsurface reservoir was not exposed to temperatures greater than roughly 180°C. Expected correlations, like for example, between grain size or sorting and porosity and permeability (e.g. Rogers and Head, 1961; Füchtbauer, 1988; Olivarius et al., 2015) could not be reproduced in this study (Fig. 2.16). This might indicate that reservoir quality cannot be predicted from granulometry in sandstones that have undergone extensive cementation, which is manifested in the very low permeability.

However, the data uncover that samples with a high content of ductile detrital grains are prone to compaction, which led to the development of a pseudomatrix (Tab. 2.1). The resulting loss of primary porosity and in permeability (e.g. Leeder and Hardman, 1990) can be attributed to early diagenesis (Fig. 2.17). This provenance-related reservoir quality control is quite apparent for sample SS16 (33 % pseudomatrix) that has the lowest porosity and permeability values (Tab. 2.1). Coarse-grained sandstones at the base of cycle sets (Fig. 2.3), deposited in mainly downstream accretion dominated braided-river beds, tend to have minor amounts of pseudomatrix. Therefore, they may potentially feature better reservoir quality conditions if unaffected by high temperatures. However, in the case of the studied rocks, this coarse-grained lithofacies, in turn, tends to have the highest quartz cementation (Tab. 2.1). Uncemented illite meshwork from dissolved feldspars forming secondary microporosity occur mainly in medium- to fine-grained sandstones and small pores of coarse-grained sandstones (Tab. 2.1). Framework grain dissolution and related clay authigenesis, for example kaolinite formation, led to an early gain of (micro) porosity.

While undifferentiated secondary pore space reflects the dissolution of unstable alumosilicates, rhombic-shaped pores are the consequence of dissolved euhedral ankerite crystals. The absence of compaction of these pores as well as the lack of late quartz cements and the precipitation of iron (oxide-)hydroxide within them can be linked to a telogenetic process. The observed present day porosity should have been increased by telogenetic feldspar and carbonate dissolution since the Early Tertiary after the Late Cretaceous inversion. However, this 50 Ma did not affect the tight sandstones significantly since a small amount of detrital feldspar (Tab. 2.1) is still present in the rock. The tight sandstones and intercalated shales and coals thus provided a sufficient seal against telogenetic alteration. Therefore, the final Quaternary uplift to the surface eventually allowed surface waters to interact with the rock matrix, enabling the dissolution processes. With this, the generation of this secondary porosity was deemed irrelevant to subsurface reservoir quality.

2.6.3.1 Structural control on diagenesis and porosity/permeability

Certain areas close to faults (Tab. 2.1) exhibit higher quartz cementation. However, the authigenic quartz volume varies and the data do not reflect a trend of increasing authigenic quartz volume with decreasing distance from the faults (cf. Tab. 2.1). Hence, a correlation of increased matrix quartz cements with distance to the fault and fault-related quartz veins cannot be drawn from the given data.

The lateral porosity profiles indicate that matrix porosity and permeability is higher around the W-E striking faults (Fig. 2.10). Increased porosity of up to 26.3 % (Fig. 2.10) and permeability of up to 105 mD (Fig. 2.8) can be observed in 3 – 5 m wide corridors around faults. This can be accounted for by overall minor amounts or absence of pseudomatrix, lithic fragments and quartz-cemented illite meshes (Fig. 2.18) besides carbonate and feldspar dissolution. The high porosity and permeability values might be a result of leaching- or wash-out processes, which may be based on focused fluid flow. An ascending fluid flow during mesodiagenesis would represent a geochemically open system (Bjørlykke and Jahren, 2012). Fault-related W-E striking extension veins and the inversion-related, high-saline and CO_2-bearing fluids (q.v. second quartz vein generation in Chapter 3.5.4) allude to such ascending fluids along the fault's damage zone. This is further evidenced by the mineralization of iron, lead and zinc in quartz veins and unconformably overlying Zechstein carbonates in the studied quarry (e.g. Lotze, 1953; Voigt, 1960). However, an indicative ore mineralization, equivalent to vein mineralization, is absent in the open pore space. Therefore, the high-porosity zones around the faults could not be related to a subsurface relevant fluid flow event based on the available data.

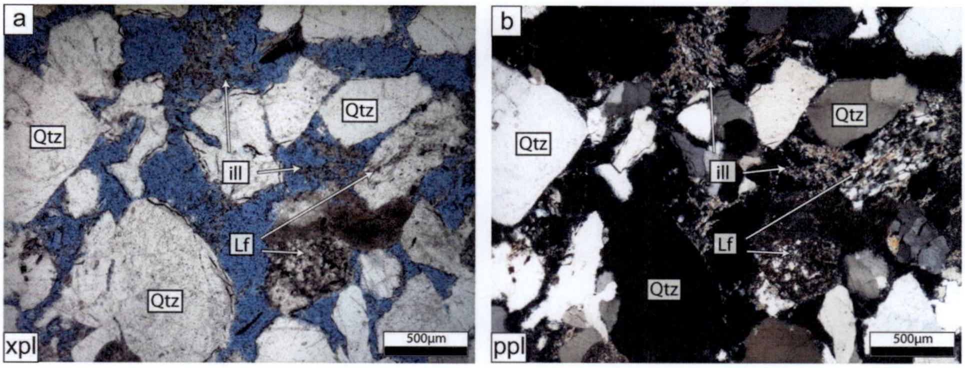

Figure 2.18: Well-connected open pore space of an extensively leached rock sample, which was taken directly out of a fault zone of a W-E striking fault with an offset > 15 m (cf. Figs. 2.4, 2.11). The matrix porosity (based on helium pycnometry) of the sample is 26.3 % and the permeability is 105 mD. The overall minor amount or absence of pseudomatrix, lithic fragments and illite meshes associated with an intense quartz cementation contribute to the increased porosity and permeability of fault zone samples (cf. Tab. 2.1). [Qtz = detrital quartz, Lf = lithic fragment, ill= authigenic illite].

2.6.4 Analog studies in reservoir evaluation

The Piesberg is an excellent high-resolution outcrop analog to tight gas reservoirs of Central Europe with respect to structural inventory, sedimentology, provenance as well as principal detrital composition. The documented main orientations of fault strikes in the quarry are in accordance with fault strikes of some of the producing Upper Carboniferous tight gas fields in the Lower Saxony Basin, 50 to 100 km further north (Fig. 2.1) (Wimmers and Koehrer, 2014). Therefore, the fault architecture reflects that of subsurface reservoirs. The fourth-order cyclicity, sedimentary structure and stratigraphic age is comparable to interpreted well data from subsurface tight gas sandstone reservoirs (Wimmers and Koehrer, 2014). However, actual reservoirs at depth lack a telogenetic influence and should provide lower porosity as along with permeability data.

While the porosity of our analog matches actual reservoir data, its paragenetic formation does not reflect reservoir conditions. A diagenetic study is indispensable for linking the timing of pore space formation with petrophysical data of analogs and actual reservoirs. Nevertheless, the lateral porosity variation in unfaulted zones might indicate subsurface trends as they reflect the natural heterogeneity of fluvial sandstones (e.g. McKinley et al., 2011).

The temperature overprint of the studied tight siliciclastics and the associated diagenetic overprint results in permeability data one to two orders of magnitude below that of actual tight reservoirs (Fig. 2.8). This demonstrates the exploration risk of tight reservoirs either because of locally higher overburden (e.g. Senglaub et al., 2006; Bruns et al., 2013) or similar structural positions. Consequently, the Piesberg analog only partially serves as an analog for Carboniferous tight gas fields in Central Europe.

2.7 Conclusions

Reservoir characteristics of tight fluvial sandstones within an Upper Carboniferous outcrop analog in Northern Germany were investigated by integrating petrophysical, structural, stratigraphic and diagenetic data. The quarry exposes fluvial sedimentary fining-upward cycles dominated by coarse- to fine-grained sandstone bodies with a net sand volume of approximately 80 %. The sedimentology and dimension of both the outcrop as well as sedimentary bodies match and reflect reservoirs at depth. The study area is located at the intersection of two different normal fault strike directions, well reflecting the structure of some of the producing Upper Carboniferous tight gas fields in the North German Basin. W-E striking faults and graben structures have throws between 2 and 20 m and NNW-SSE striking normal faults have throws up to 600 m.

Strong compaction during eodiagenesis led to an early loss of primary porosity within the sandstones. The porosity in the outcrop analog matches reservoir porosities and is laterally consistent in single horizons over 56 m, ranging between 4.5 to 8.7 %. However, the porosity was lower than in actual reservoirs during burial. Present day permeability in the outcrop analog is still one to two orders of magnitudes lower compared to reservoir data. This is the result of a high-temperature event during burial and can be associated with mesogenetic intense quartz cementation and illitization. This is well-documented by quartz-cemented mesogenetic meshwork illite, which was impregnated with bitumen based on hydrocarbon charging. The correlation of petrographic and petrophysical data shows that expected facies-related reservoir quality trends were overprinted by high-temperature diagenesis. The telogenetic dissolution of carbonates and unstable alumosilicates mainly contributed to present day macroporosity. Matrix porosity is strongly enhanced in fault corridors of W-E striking faults to porosities of up to 26.3 % and permeabilities of up to 105 mD, stemming from the leaching processes because of focused fluid flow in the fault damage zones. However, a telogenetic origin cannot be ruled out. The outcrop is a reservoir analog for detrital and authigenic composition within sedimentary bodies. It reflects relative porosity and permeability trends and possibly their spatial heterogeneity. However, the outcrop is not a reservoir analog with respect to the intensity of diagenesis, resulting from a higher temperature and telogenetic overprint. Thus, the outcrop does not reflect absolute porosity and permeability values and the unusually low permeability may represent a reservoir risk for tight gas exploration.

3 Kilometer-scale fault-related thermal anomalies in tight gas sandstones

3.1 Abstract

Upper Carboniferous sandstones make one of the most important tight gas reservoirs in Central Europe. This study integrates a variety of geothermometers (chlorite thermometry, fluid inclusion microthermometry and vitrinite reflection measurements) to characterize a thermal anomaly in a reservoir outcrop analog (Piesberg quarry, Lower Saxony Basin), which is assumed responsible for high temperatures of circa 300°C, deteriorating reservoir quality entirely. The tight gas siliciclastics were overprinted with temperatures approximately 90 – 120°C higher compared to outcropping rocks of a similar stratigraphic position some 15 km to the west. The local temperature increase can be explained by circulating hydrothermal fluids along the fault damage zone of a large NNW-SSE striking fault with a displacement of up to 600 m in the east of the quarry, laterally heating up the entire exposed tight gas sandstones. The kilometer-scale lateral extent of this fault-bound thermal anomaly is evidenced by vitrinite reflectance measurements of meta-anthracite coals ($VR_{rot} \sim 4.66$) and the temperature-related diagenetic overprint. Data suggest that this thermal event and the associated highest coalification was reached prior to peak subsidence during Late Jurassic rifting (162 Ma) based on K-Ar dating of the < 2 µm fraction of the tight gas sandstones. Associated stable isotope data from fluid inclusions, hosted in a first fracture filling quartz generation (T ~ 250°C) close to lithostatic fluid pressure (P ~ 1000 bars), together with authigenic chlorite growth in mineralized extension fractures, demonstrate that coalification was not subject to significant changes during ongoing burial. This is further evidenced by the biaxial reflectance anisotropy of meta-anthracite coals. A second event of quartz vein formation took place at lower temperatures (T ~ 180°C) and lower (hydrostatic) pressure conditions (P ~ 400 bars) and can be related to basin inversion. This second quartz generation might be associated with a second event of illite growth and K-Ar ages of 96.5 – 106.7 Ma derived from the < 0.2 µm fraction of the tight gas sandstones.

This study demonstrates the exploration risk of fault-bound thermal anomalies by deteriorating entirely the reservoir quality of tight gas sandstones with respect to porosity and permeability due to the cementation with temperature-related authigenic cements. It documents that peak temperatures are not necessarily associated with peak subsidence. Consequently, these phenomena need to be considered in petroleum system models to avoid, for instance, overestimates of burial depth and reservoir quality.

3.2 Introduction

Unconventional reservoirs, such as tight gas fields, became progressively more important for exploration activities over the last decades (e.g. Law et al., 1989; Leeder and Hardman, 1990; Meckel et al., 1992; Schegg et al., 1997; Besly, 1998; Law et al., 1998, 2002; Popov et al., 2001; Kombrink et al., 2010). In North Germany and the North Sea, Upper Carboniferous tight gas sandstones host stacked reservoirs, which are sourced from the intercalated coals and sealed by intercalated claystone and Zechstein salt as top seal. Due to their low permeability, generally defined as tight below 0.6 mD (conditions for North German reservoirs; Pusch et al., 2005; Häfner, 2006; Albrecht and Reitenbach, 2015), fracture zones offer "sweet spots" for exploration (e.g. Laubach et al., 2009), which are likely expected around faults (e.g. Faulkner et al., 2010; Torabi and Berg, 2011).

Thermal anomalies are well known from many sedimentary basins, caused, for example, by varying thermal conductivities of salt domes (e.g. North German Basin or the Gulf of Mexico; O'Brien and Lerche, 1988; McBride et al., 1998; Magri et al., 2008) or by an increased permeability around fault zones (e.g. Cherubini et al., 2013). Fault-bound thermal anomalies are noted in several publications to be a cause for temperature increase in adjacent rocks (e.g. Bruhn et al., 1994; Andrews et al., 1996; Curewitz and Karson, 1997; Fleming et al., 1998; Lampe and Person, 2000; Frings et al., 2004; Timar-Geng et al., 2004, 2009; Wisian and Blackwell, 2004; Lampe and Schwark, 2012). Zwingmann et al. (1998, 1999) associated the illitization of North German Rotliegend reservoir sandstones with hydrothermal fluid migration along large fault systems. Liewig and Clauer (2000) concluded that Jurassic illite growth based on K-Ar age determination in the same reservoir rocks can be only explained by the anomalous temperatures caused by hydrothermal activity. Will et al. (2016) observed a localized high-temperature overprint of 260 – 270°C in Upper Jurassic pelitic rocks at the Gehn complex (Lower Saxony Basin, approximately 20 km in the NNW direction of the study area) and ascribed this to fault-bound hydrothermal activity.

Although temperature anisotropies are well-known from vitrinite reflectance and fluid inclusion data from wells and outcrops in the southern part of the Lower Saxony Basin (e.g. Teichmüller, 1986; Behr et al., 1987; Sedat, 1992; Drozdzewski et al., 2009; Lüders et al., 2012; Will et al., 2016), the significance of fault-bound temperature anisotropies in tight gas reservoirs with respect to exploration risks is still unclear. Numerical petroleum system models often exclude local fault-related temperature anomalies as their lateral extent is unknown, and rather rely on simple heat flow models. Several studies, however, noted that local unexpected higher maturities in the Lower Saxony Basin may be the result of hydrothermal alteration (e.g Stadler and Teichmüller, 1971; Reutel et al., 1995; Lüders et al., 2012; Lüders and Plessen, 2015).

Our study applied a variety of geothermometers to explain the temperature anomaly of a fault-bound tight gas reservoir analog. Based on K-Ar dating, we provide evidence that the reservoir analog reached over-maturation and its final maturation stage of organic matter prior to peak subsidence due to fault-related hydrothermal fluid flow and alteration. We show that this thermal event laterally overprints tight gas sandstones over at least one kilometer.

3.3 Geological setting

3.3.1 The reservoir analog study area

The studied Piesberg quarry is an operated quarry for Upper Carboniferous sandstones and is located at the southwestern margin of the Lower Saxony Basin. The compared Woitzel quarry is located within the adjacent hard coal mining district in the Ibbenbüren Block west of the Piesberg (Fig. 3.1). The reservoir-scale Piesberg study site has dimensions of approximately 1 km in the W-E direction and 0.5 km in the N-S direction (Fig. 3.2) with a depth of roughly 140 m, enabling the investigation from the "seismic" down to the "sub-seismic" scale. The study area extends along the WNW-ESE striking Piesberg-Pyrmont axis (Drozdzewski, 1985; Drozdzewski et al., 2009) and forms a gently to the west plunging anticline (Fig. 3.2). The folding of the uplifted block was first explained by vertical movements (Hollmann et al., 1971) and, more recently, by dextral strike-slip movements along W-E directed faults (Fig. 3.2) (e.g. Drozdzewski, 1998), related to Late Cretaceous/Cenozoic inversion tectonics (e.g. Kley and Voigt, 2008; Sippel et al., 2009). At the eastern side of the quarry, a NNW-SSE striking fault with a down-dip displacement of up to 600 m (eastern normal fault, Fig. 3.2) (Hinze, 1979; Baldschuhn et al., 2001) separates the Carboniferous tight gas sandstones from the Triassic Buntsandstein. Both observed fault strike directions represent the overall structural pattern of the region and the reservoirs further north (Fig. 3.1).

The fluvial cyclic deposits of the Westphalian C and D (Pennsylvanian) exposed in the studied quarry are dominated by clastic sedimentary rocks with intercalated coal seams (Fig. 3.3). A major coal seam in the studied quarry (locally called "Zweibänke") marks the boundary between the Westphalian C and D (c.f. Köwing and Rabitz, 2005) and can be correlated with the active hard coal mining district of Ibbenbüren (Fig. 3.1) (i.e.

"Dickenberg" coal seam; Köwing and Rabitz, 2005). The "Bänkchen" coal seam in the Piesberg quarry can be correlated with the "Alexander" coal seam of the Woitzel quarry (e.g. David, 1990; Jones and Glover, 2005), which is located approximately 15 km to the west of the Piesberg (Fig. 3.1). Rotliegend siliciclastics occur in central parts of the Lower Saxony Basin but are absent in the study area, where the Zechstein unconformably overlies the Upper Carboniferous (Fig. 3.3) (Voigt, 1960; Fiedler, 1984). Zechstein carbonates are metasomatically transformed to siderite and brown iron ore at that interface (Lotze, 1953; Voigt, 1960) or are replaced by zinc-(lead) sulfides which have been mined for several decades at the mine Perm (e.g. Röhrs, 1992).

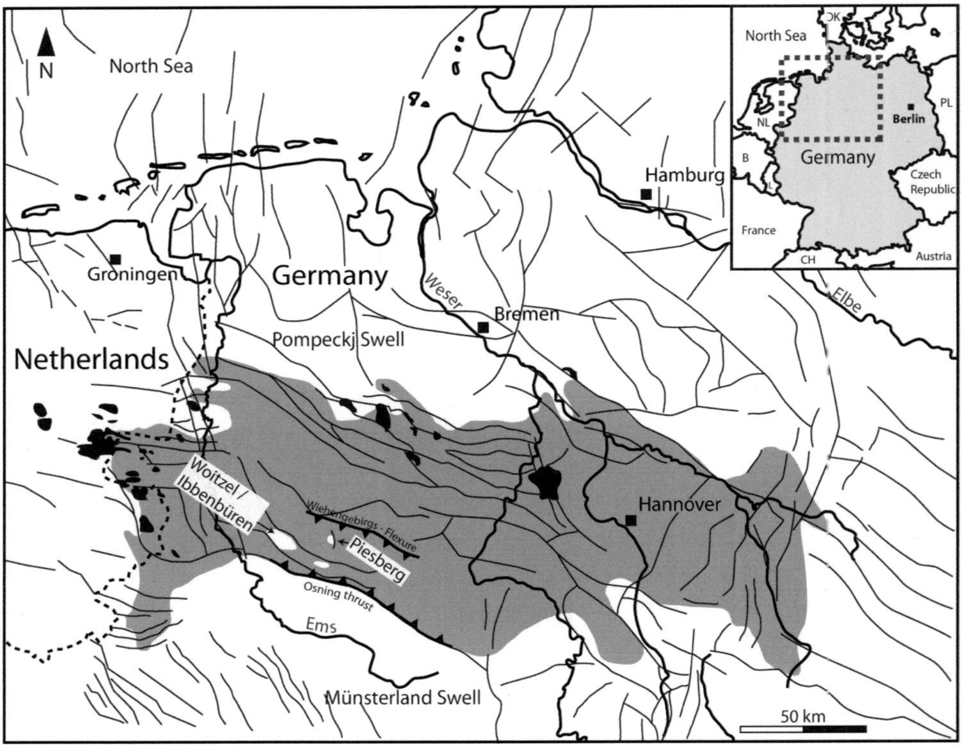

Figure 3.1: Uplifted Upper Carboniferous blocks (white) in the hanging wall of the Osning thrust in the south of the Lower Saxony Basin (grey area, based on Senglaub et al., 2006). The Woitzel quarry is hosted in the Ibbenbüren hard coal mining district. The basement fault block pattern of the region (dark grey solid lines) is based on Baldschuhn and Kockel (1999) and extended with fault data from Kombrink et al. (2010). Black areas represent actual subsurface Carboniferous gas fields (after Kombrink et al., 2010). The dotted black line represents the border between Germany and the Netherlands. Solid black lines indicate the coastal line as well as rivers (Ems, Weser, and Elbe).

Structurally, the study area and the Ibbenbüren mining district are located in the hanging wall of the Osning thrust (Fig. 3.1), a system of normal faults that have been reactivated to form thrust faults in the Late Cretaceous (e.g. Baldschuhn and Kockel, 1999; Kley et al., 2008). The late Cretaceous inversion is one of the controlling factors involved in the formation or modification of hydrocarbon traps in Carboniferous reservoirs in NW Germany and the North Sea (Kombrink et al., 2010). This is reflected, for example, by the uplifted Carboniferous blocks in the study area and the Ibbenbüren hard coal mining district, hydrocarbon reservoir blocks Husum-Schneeren and Rehden and elsewhere (Hollmann et al., 1998; Drozdzewski et al., 2009).

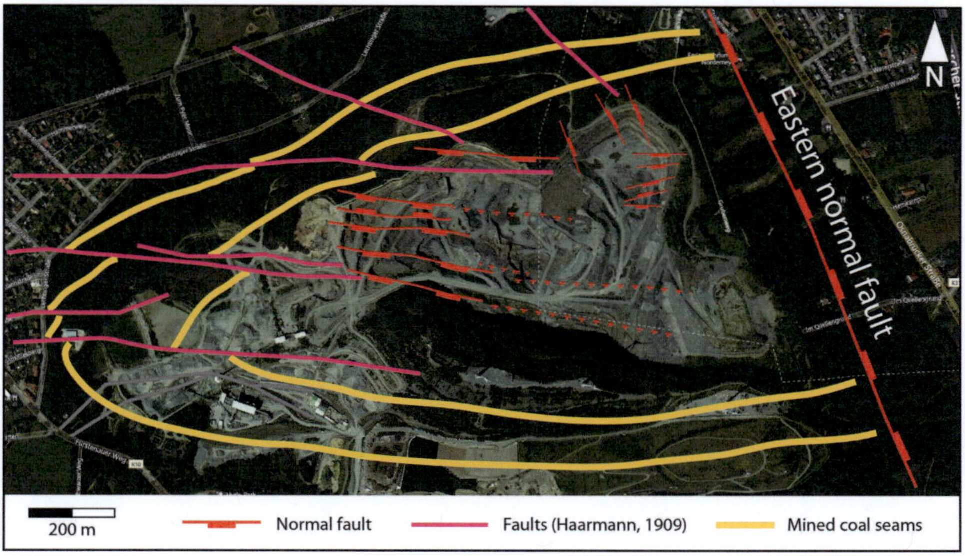

Figure 3.2: Aerial image of the Piesberg quarry (Bing Maps, 2015). The coal seams and faults (pink), including the major normal fault in the east, are based on the historical subsurface mining data of Haarmann (1909). Normal faults (red) with dip direction indicators were observed within this study.

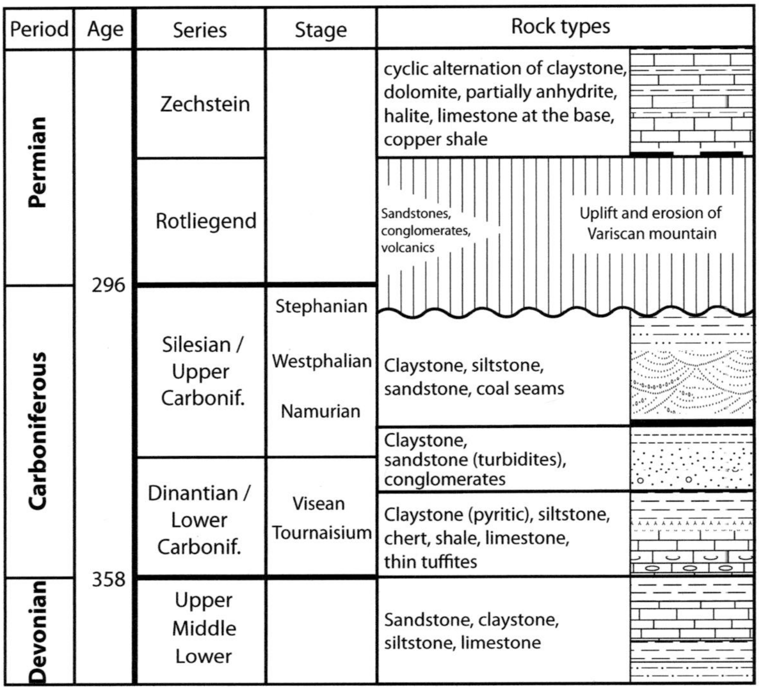

Figure 3.3: Stratigraphy of the Piesberg region (after Drozdzewski and Ribbert, 2003).

3.3.2 Basin evolution and thermal models

Basin evolution started with sedimentation in the foreland basin of the northwestwards prograding Variscan orogeny in the Late Carboniferous southeast of the study area (Ziegler, 1977; Glennie, 1986; Brink et al., 1992; Franke, 1995). Large fluvial systems interfering with the occurrence of swamp areas dominated the region (e.g. Drozdzewski et al., 2008). The resulting fluvial cyclic deposits of the Westphalian C and D (Pennsylvanian) exposed in the studied quarry are dominated by clastic sedimentary rocks with intercalated coal seams. The deposition of some organic-rich material began already during the Namurian as indicated by the presence of minor coal seams. In Upper Carboniferous times, the sedimentary conditions changed from a marine to a (partially) terrestrial depositional environment (e.g. Hedemann et al., 1984) in the Westphalian with abundant coal seams (Scheidt and Littke, 1989).

Minor subsidence in the Permian was associated with thermal lithosphere relaxation as well as sedimentation (e.g. Scheck and Bayer, 1999; van Wees et al., 2000). Triassic subsidence was controlled by rifting because of extension, which continued until the Late Cretaceous (e.g. Brink et al., 1992; Senglaub et al., 2005; Lohr et al., 2007; Kley and Voigt,

2008). The uplift of the Upper Carboniferous, with the onset of inversion tectonics in the Late Cretaceous, led to the exposure of the studied rocks at the surface since the Miocene (e.g. Senglaub et al., 2005).

The southern rim of the Lower Saxony Basin is characterized by highest thermal maturity of organic matter (e.g. Teichmüller et al., 1984; Teichmüller, 1986; Petmecky et al., 1999; Drozdzewski et al., 2009), with its peaks in the Wiehengebirge flexure zone (Fig. 3.1). The reason for this high maturity is still widely debated. Some authors suggested the intrusion of a deep-seated pluton during the Early Cretaceous (referred to as the "Bramsche Massif") to explain the anomalous maturity in the basin center (e.g. Bartenstein et al., 1971; Stadler and Teichmüller, 1971; Teichmüller et al., 1984; Betz et al., 1987; Brink, 2013). This model was, amongst others, based on magnetic and gravimetric anomalies centered at the proposed localities of the magmatic bodies (e.g. von Flotow et al., 1931; Reich, 1933; Breyer, 1971; Bachmann and Grosse, 1989).

On the other hand, the existence of such an intrusive body was doubted at the beginning of this century and deep subsidence models were suggested in order to explain highest temperatures at peak subsidence during the Late Cretaceous (e.g. Petmecky et al., 1999; Senglaub et al., 2006; Brink, 2013; Bruns et al., 2013). The deep burial scenarios explaining high maturity trends imply high heat flows during the Late Jurassic and Early Cretaceous with erosion rates of 6700 m (e.g. Bruns et al., 2013) following basin inversion and resulting uplift in Late Cretaceous/Cenozoic times. Gaupp and Okkerman (2011) noted that maximum temperatures were reached before maximum burial for Permian sandstones in the Netherlands.

3.4 Methodology

Petrographic investigation was performed on thin sections with stained porosity by blue epoxy resin. The petrographic descriptions were completed with analyses on selected samples by scanning electron microscopy using a Zeiss ULTRA PLUS FE-SEM equipped with a secondary electron (SE) and a back-scatter electron (BSE) detector as well as an energy dispersive spectrometer system (EDX). The measurements were carried out with an accelerating voltage of 15 kV and a working distance of approximately 10 mm.

Chlorite elemental composition measured on polished thin sections with an electron microprobe (JEOL Superprobe JXA-8900R; measurement conditions: accelerating voltage of 15 kV, beam current of 24 mA, beam diameter of 10 μm, ZAF-correction) was employed for geothermometer calculations as tetrahedral and octahedral occupancies in chlorites are indicative for the temperature of formation (Jahren and Aagaard, 1989). However, Essene and Peacor (1995) and Merriman and Peacor (1999) warned that all chlorite thermometers need to be applied with caution. Empirical approaches, in which temperatures are directly derived from chlorite compositions, are inappropriate and can be applied only to certain calibrated and restricted geological settings (de Caritat et al., 1993; Bourdelle et al., 2013). Lanari et al. (2014) addressed these limitations and introduced a thermodynamic activity-composition model for di-trioctahedral chlorite. Their semi-empirical geothermometers cover a wider range of temperatures than previous calibrations. The temperature calculation was based on equation 40 of Lanari et al. (2014) and a pressure of 100 MPa was assumed. Only chlorite analyses with 84.5 to 87 oxide wt. % were used for further calculations and recalculated based on 14 oxygen equivalents. Additionally, measurements not representing elemental composition of chlorite were excluded. These might result from mixed analysis with other clay components of the sample.

For K-Ar age determination, the < 2 µm fraction was separated from material of four samples with the Atterberg method after mechanical milling with a McCrone micronizing mill for 15 minutes with ethanol. Additionally, the < 0.2 µm fraction was separated from the same samples by centrifuging the milled sample material for 32 minutes at 2400 rpm and subsequent freeze-drying. For all four samples, the < 2 µm fraction was extractable, while two samples did not produce a < 0.2 µm fraction. The argon isotopic composition for the K-Ar age calculation for all fractions was determined by the isotope dilution method using a highly enriched ^{38}Ar spike and a Thermo Scientific ARGUS VI ™ noble gas mass spectrometer. Potassium content was determined in duplicates with a BWB-XP ™ flame photometer. Age determinations are based on constants recommended by the International Union of Geological Sciences as quoted in Steiger and Jäger (1977). Analytical errors of the age calculations were according to 95% confidence intervals.

Four samples of anthracite coals were collected from three different coal seams across the quarry for vitrinite measurements. The samples were embedded in epoxy resin and hardened at 37°C in order to prepare polished sections with a Struers Tegra Pol21 polishing system combined with a Tegra Force 5 head. The grinding/polishing process comprised different stages to obtain a final surface roughness of < 1 µm after the final polish. The reflectance analyses were conducted with a Zeiss Axio Imager.M2M microscope with a Zeiss VIS-LED illumination and a Basler Scout camera. For the analyses, a 50x/1.0 Epiplan-NEOFLUAR oil immersion objective was used as well as a cubic zirconia mineral standard with 3.125% reflectance to calibrate before each measurement. The rotational vitrinite reflectance was determined via an automated polarization filter with 15° increments in the polarization angle interval between 0° and 180° (q.v. Bruns and Littke, 2015). For data acquisition and processing, an automated software package was utilized (DISKUS Fossil, Technisches Büro Carl H. Hilgers). Reflectance measurements were taken on at least 100 particles per sample. The microscope-specific partial polarization was corrected by calibration with the isotropic mineral standard (cubic zirconia) and deriving a correction factor array for each sample over the whole polarization interval (e.g. Bruns and Littke, 2015).

Fluid inclusions were studied in doubly polished thick sections of quartz from partially and completely sealed veins by conventional microthermometry using a FLUID INC.- adapted USGS heating/freezing system mounted on a BX50 Olympus microscope with a 40x objective (N.A. = 0.55). The heating/freezing stage was calibrated with synthetic fluid inclusion standards supplied by Synflinc. Gas-bearing fluid inclusions in quartz were analyzed for molecular compositions with a Jobin-Yvor LabRam confocal Raman microspectrometer equipped with a 532.6 nm Nd-YAG laser (100mW) and a SLMPlan 100x/0.90 objective lens (Olympus). Silicon (520 cm^{-1}) and diamond (1332 cm^{-1}) were used for internal wavenumber/spectral calibration. The Raman instrument factor was determined with synthetic inclusions from Virginia Tech. Raman spectra of gaseous inclusions were collected in the spectral range between 1200 and 3000 cm^{-1} with a Peltier cooled collector. A 3 x 30 s acquisition time was used for CH_4-CO_2-bearing inclusions in quartz I. CO_2-CH_4-N_2- bearing inclusions in quartz II were measured at 3 x 60 s in order to improve the signal-to-noise ratio.

Carbon stable isotope analyses of fluid inclusion gases were performed using a sample crusher connected via a gas chromatography (GC) column to an element analyzer (EA)-IRMS system. The isotopic compositions of $\delta^{15}N$ of N_2 (when present), $\delta^{13}C$ of CO_2 (from oxidation of CH_4) and $\delta^{13}C$ of natural CO_2 from fluid inclusions were measured online simultaneously using reference gases calibrated against IAEA-N1 for N_2 and NBS19 for CO_2. For details, see Plessen and Lüders (2012).

Matrix porosity on plugs from the quarry was determined with helium pycnometry by a non-commercial device as described by Ghanizadeh et al. (2014). The measuring circuit, including sample and reference cells, was tempered to 30°C. The measurements were carried out on plugs dried at least overnight (105°C, vacuum) and porosities were derived from skeletal and bulk volumes. Permeability of plug samples from the quarry was measured under confining fluid pressure (30 MPa) by a steady-state technique as described by Ghanizadeh et al. (2013) with a non-commercial device based on a hydrostatic pressure cell. A Klinkenberg correction was applied to measured gas permeability (helium) to account for slip-flow effects.

3.5 Results

3.5.1 Structural inventory

3.5.1.1 Faults

Within the Piesberg quarry, two dominant fault strike directions can be distinguished. The NNW-SSE striking orientation is represented with the most prominent fault in the east of the quarry (Fig. 3.2, not exposed anymore), featuring a down-dip displacement of the block east of the fault of up to 600 m (e.g. Hinze, 1979; Baldschuhn et al., 2001). In the NE of the quarry, this strike direction is represented by several minor faults still exposed with a throw of up to 5 m (Fig. 3.2). The second main strike orientation of the normal faults is W-E to WNW-ESE with fault displacements between 2 m and more than 15 m. Younger strike-slip slickenlines truncate older dip-slip slickensides, indicating a change in the fault displacement direction (Wüstefeld et al., 2014).

3.5.1.2 Fracture-fill mineralization

Quartz-cemented veins may be both partially and completely sealed. At least two generations of quartz can be discriminated. An older quartz (quartz I) cemented fracture mineralization is generally entirely sealed and microstructures exhibit syntaxial growth (q.v. Becker et al., 2011) with locally younger euhedral crystals (quartz II) in the center of the vein (Fig. 3.4a). These quartz I veins contain chlorite flakes within the vein and along the vein-wall interface (Fig. 3.4a). Other exposed subvertical quartz veins are oriented subparallel to the W-E striking faults and are restricted to the vicinity of faults with a larger offset (> 10 m). Around a fault with approximately 15 m offset, the veins occur in a 36 m wide corridor, restricted to a competent sandstone bed, their spacing increasing with greater distances from the fault. Veins distal from the fault are entirely sealed with quartz, while partially filled veins with elongated quartz crystals (quartz II) are more prominent along the fault. Microstructures within quartz veins show syntaxial growth, while others and vein tips feature an ataxial microstructure (e.g. Becker et al., 2011). Quartz II veins commonly are associated and/or inter-grown with ore minerals, such as sphalerite and marcasite, the latter often altered to iron (oxy-)hydroxides. This is in agreement with the reported ore deposits in the region at the Piesberg (Lotze, 1953; Voigt, 1960).

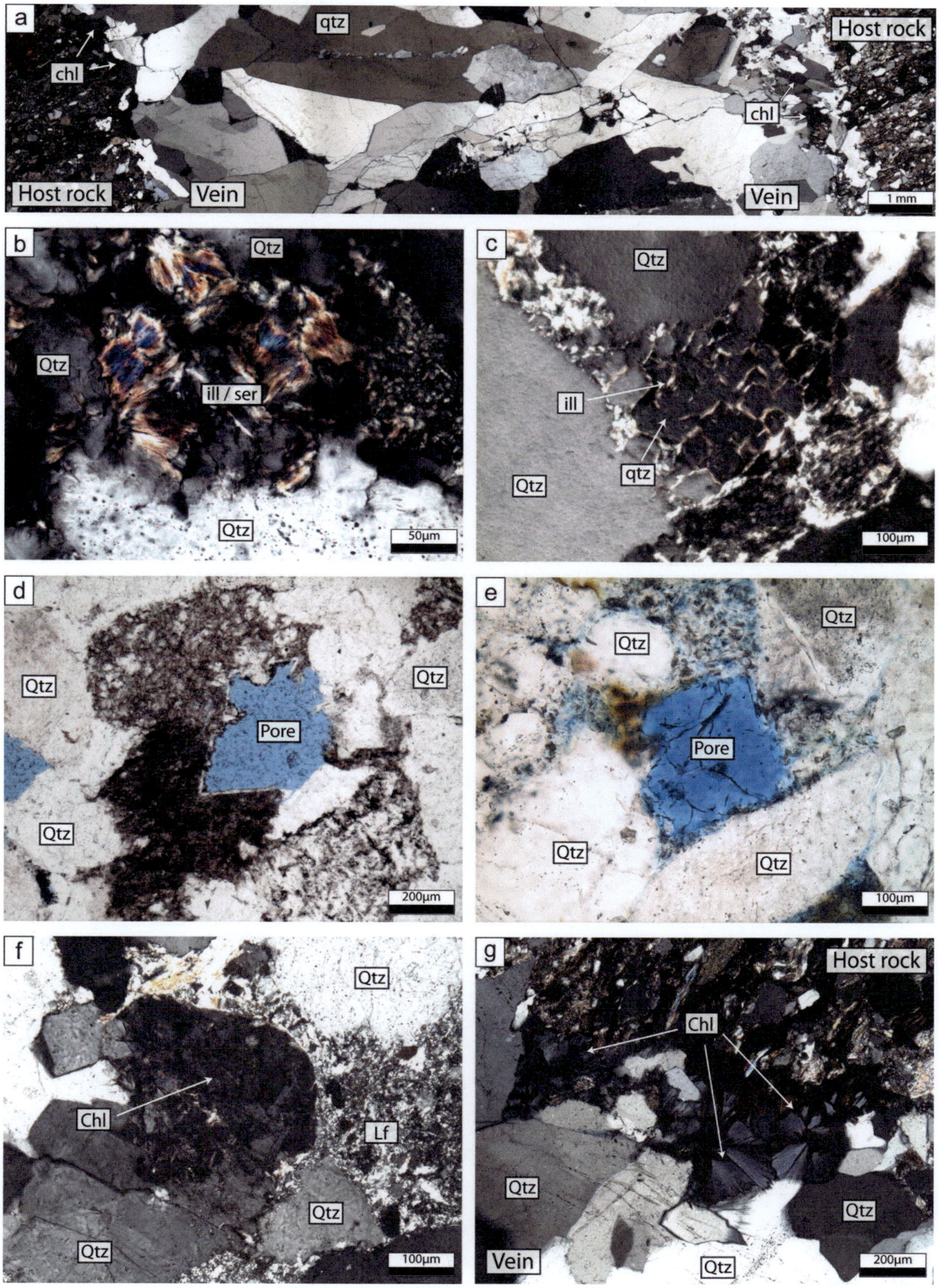

Figure 3.4 (Description on next page)

Figure 3.4 (Previous page): a) Sealed quartz (qtz) vein cement exhibiting syntaxial growth with euhedral crystals in the center with flakes of chlorite (chl) enhosted within the quartz vein as well as at the vein-wall interface. b) Illitized kaolinite booklets with unusual high interference colors indicating a subsequent sericitization. c) Quartz (qtz) cemented fibrous meshwork illite (ill) grown in secondary pore space between detrital quartz grains (Qtz). d) & e) Rhombic secondary pores as a result of dissolved (iron-rich) carbonates in the Piesberg quarry (e) and in the Woitzel quarry (Ibbenbüren hard coal mining district) (d). f) Pore-filling chlorite (chl) within former detrital clasts show blocky to pseudohexagonal shapes. g) Chlorite (chl) inter-grown with quartz (qtz) in vein cements feature well-developed fan-shaped aggregates. Both pore- and vein-filling chlorite show anomalous blue interference colors in polarized light.
[Chl = chlorite; ill = illite; qtz = quartz; ser = sericitized illite; lf = lithic fragment]

3.5.2 Reservoir properties and diagenetic evolution in the region

The present day matrix porosity and permeability of the study area reveals lower values compared to the Woitzel quarry (Fig. 3.5) at the same correlative stratigraphic position approximately 15 km in the western direction (Fig. 3.1). Piesberg data features an average (arithmetic mean) porosity of 7 % (ranging between 0.5 – 16 %) and mean (geometric) permeability of 0.0003 mD (confining pressure of 30 MPa, ranging between 0.00004 and 0.03 mD) (Fig. 3.5). In contrast to the majority of the Piesberg sandstones, Woitzel samples from the same stratigraphic unit have better reservoir properties, with an average porosity of 16 % (ranging between 13 and 23 %) and mean permeability of 0.03 mD (confining pressure of 30 MPa, ranging between 0.01 and 1 mD) (Fig. 3.5).

The exposed Upper Carboniferous strata in both quarries comprise a detrital composition of predominantly quartz (mono, poly- and microcrystalline), rock fragments and minor amounts of feldspar, which can be classified as (sub-)litharenite according to McBride (1963). Kaolinite is a common authigenic clay mineral and generally underwent the process of illitization during burial, which is characterized by the replacement of individual kaolinite platelets to illite (Fig. 3.4b). Being the most abundant clay mineral, authigenic illite is also an alteration product of detrital K-feldspars as well as rock fragments (Fig. 3.4c). Both sandstones show evidence of former (iron-rich) carbonate cementation, which was dissolved due to fluid-rock interactions with surface waters during the latest stages of the diagenetic evolution. This is recorded by the presence of characteristic rhombic pore shapes partially filled with iron (oxide-)hydroxides (Fig. 3.4d, e), which is the origin of most of the observable present day macroporosity.

The general evolution of matrix permeability and porosity in Upper Carboniferous sandstones of both localities is controlled by several factors. Sandstone porosity is destroyed by mechanical and chemical compaction. Pore space was occluded by pseudomatrix formation during early subsidence (e.g. Bertier et al., 2008; Wüstefeld et al., 2014), which is also associated with the process of mechanical compaction. While being present in the Piesberg quarry, high-temperature mineral assemblages, such as chlorite, are absent in the Woitzel quarry (Fig. 3.4f, g). Sedimentary rocks of the Piesberg quarry feature a generally stronger illite authigenesis and subsequent sericitization of the same, manifested in high interference colors, which are rather typical for muscovite than for illite (Fig. 3.4b). Besides a higher content of authigenic quartz in the Piesberg quarry, fibrous authigenic illite meshwork is commonly entirely cemented with quartz, which cannot be observed in Woitzel sandstones (Fig. 3.4c).

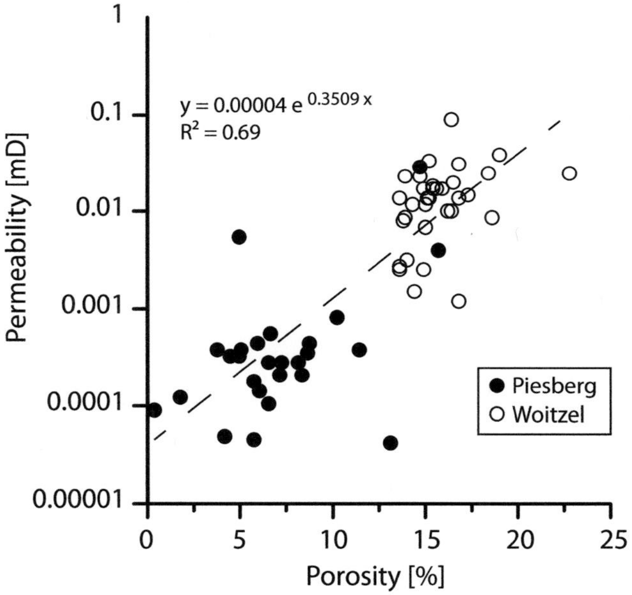

Figure 3.5: Permeability-porosity cross-plot for the study area and an Upper Carboniferous outcrop of a same correlative stratigraphic position approximately 15 km in the western direction (Ibbenbüren coal district). All data are given for 30 MPa confining pressure. Further details on Woitzel data in Becker et al. (2017).

3.5.3 Chlorite thermometry

Chlorite occurs as flakes and as well-developed fan-shaped aggregates inter-grown with quartz in vein cements (Fig. 3.4g) and as pore-filling chlorite replacing detrital grains (Fig. 3.4f). The latter is characterized by blocky to pseudohexagonal shape (Fig. 3.4f). In the thin sections, both types are grey to pale green and exhibit first-order anomalous blue interference colors in polarized light (Fig. 3.4f,g).

Electron microprobe analyses revealed that the pore-filling chlorite within former detrital clasts is composed of the IIb (β= 97°) polytype (Curtis, 1985) with average contents of tetrahedral Al^{3+} of 2.82 per formula unit (p.f.u.) and Fe/(Fe+Mg) ratios of 0.70 on average (n = 128) (Fig. 3.6). The composition can be classified as chamosite (Curtis, 1985). The fan-shaped chlorite within vein cements are also composed of the IIb polytype (Fig. 3.6). The average contents of tetrahedral Al^{3+} are 2.81 p.f.u. with average Fe/(Fe+Mg) ratios of 0.67 (n = 35). This chlorite (also chamosite) has higher contents of MgO compared to the pore-filling chlorite (8.46 wt. % vs 7.54 wt. %). The pore-filling chlorite has average model temperatures of 270°C whereas the chlorite in veins has average model temperatures above 300°C (Fig. 3.7, cf. Lanari et al., 2014).

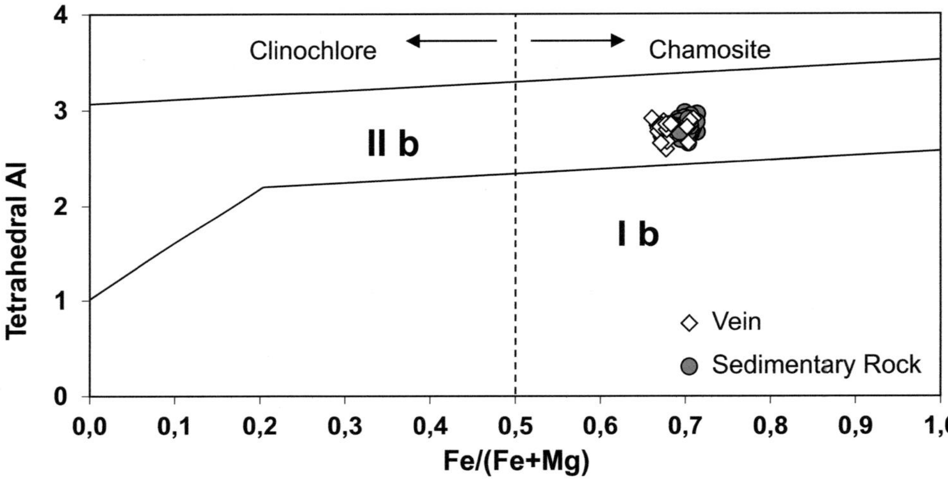

Figure 3.6: Tetrahedral Al vs. Fe/(Fe+Mg) ratio of both chlorite varieties (after Curtis et al., 1985). The chlorites plot within the IIb domain (high-T modification). All investigated chlorites can be classified as chamosite based on the Fe/(Fe+Mg) ratio (n = 163).

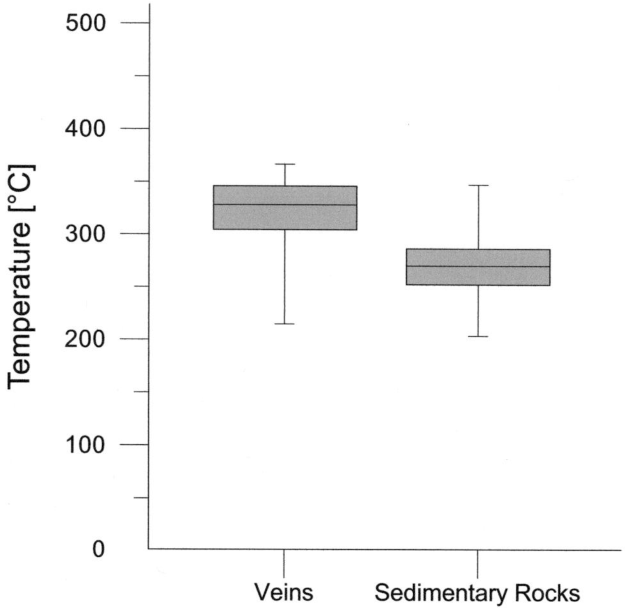

Figure 3.7: Box whisker plot of temperatures derived by chlorite thermometry. Temperatures of pore-filling chlorite (n = 128) are slightly lower than of vein cement (n = 35).

3.5.4 Fluid inclusion analyses

Fluid inclusions were studied in samples of two generations of quartz vein cement, hereinafter named quartz I and II, respectively. While quartz I samples are characterized by the presence of chlorite at the vein-host rock interface of the quartz vein (Fig. 3.4a, g), it is absent in quartz II. Quartz II is often inter-grown with (iron) ore minerals, like marcasite, which is not observed for quartz I. Both generations of quartz host two compositional types of fluid inclusions, namely aqueous two-phase and single-phase gas-bearing inclusions. Primary aqueous two-phase inclusions are common in quartz I and often occur in well-defined growth zones together with gas-rich inclusions (Fig. 3.8a), suggesting co-genetic entrapment within an immiscible system. Heterogeneous trapping of fluid and gas inclusions is supported by the presence of traces of gases in the vapor bubbles of some aqueous two-phase inclusions detected by Raman spectroscopy. Gas-rich inclusions in quartz II are highly frequent whereas aqueous two-phase inclusions are less abundant. Secondary fluid inclusions in quartz II often show a negative-crystal shape and are arranged along planes and trails (Fig. 3.8b), suggesting that they formed by healing of cracks (necking-down). In some samples, primary gas-rich and aqueous two-phase inclusions in quartz II decorate rare growth zones or occur together in isolated fluid inclusion assemblages. However, it is noteworthy that secondary fluid and gas inclusions formed by healing of cracks in individual quartz II samples exhibit the same ranges of homogenization temperatures and salinities as well as molecular compositions, respectively, as primary fluid and gas inclusions from the same sample. This may be indicative for healing of cracks during crystal growth.

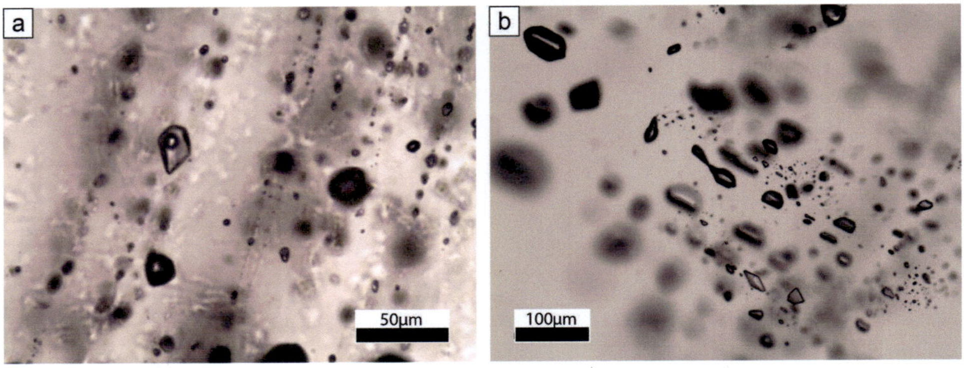

Figure 3.8: a) Primary or pseudosecondary aqueous two-phase and gas-rich inclusions decorating a growth zone in quartz I. b) Healed crack in quartz II with secondary gas-rich inclusions individually showing a negative crystal shape due to "necking-down".

Fluid inclusions in quartz I and II differ from each other in terms of salinity, homogenization temperatures and gas composition. Melting temperatures of ice in aqueous two-phase inclusions in quartz I point to salinities in the range of 13.8 – 15.7 wt. % NaCl equiv. whereas aqueous two-phase inclusions hosted in quartz II have higher salinities between 17.7 and 21.6 wt. % NaCl equiv. The homogenization temperatures of aqueous two-phase inclusions hosted in nine studied samples of quartz I range between 220 and 260°C (mean 251°C) and are considerably higher than the homogenization temperatures in the seven studied quartz II samples (167 – 202°C, mean 182°C). Gas-rich inclusions in quartz I consist of mixtures of CH_4 and CO_2 (Fig. 3.9) and exhibit melting of solid CO_2 followed by homogenization of CH_4 to the liquid phase (V→L) upon freezing. The observed low-temperature phase transitions of CH_4-CO_2-bearing inclusions in quartz I point to molar volumes between 48 and 62 cm^3/mol in the CH_4-CO_2 system (Thiery et al., 1994). In contrast, the studied gas-rich inclusions in quartz II comprise of mixtures of CH_4, CO_2 and N_2 (Fig. 3.9). During freezing runs, the studied gas-rich inclusions in quartz II always show melting of solid CO_2 and homogenization of the vapor bubble into the vapor phase (L→V). Molar volumes between 95 and 105 cm^3/mol for CH_4-CO_2-N_2 inclusions in quartz II were calculated using the software of Bakker (2003).

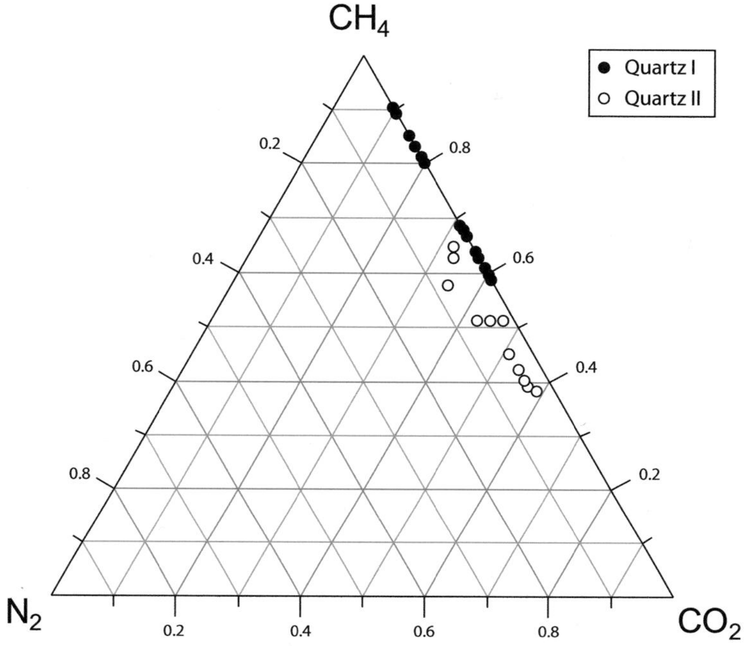

Figure 3.9: Composition of gas-rich fluid inclusions in quartz I and II.

Stable isotope bulk analyses of fluid inclusions in small chips of quartz I and II show $\delta^{13}C_{(CH4)}$ values that do not significantly differ from each other (Fig. 3.10). They are clearly enriched in ^{13}C when compared with the $\delta^{13}C_{(CH4)}$ values of fluid inclusions hosted in quartz veins from the Ibbenbüren coal mine approximately 15 km to the west of the Piesberg (Figs. 3.1, 3.10). While $\delta^{13}C$ (CO_2) values are positive for fluid inclusions hosted in quartz I (2.5 – 6.5 ‰ VPDB), $\delta^{13}C$ (CO_2) values of fluid inclusions hosted in quartz II are slightly negative to positive (-2.5 – 6 ‰ VPDB) (Fig. 3.10).

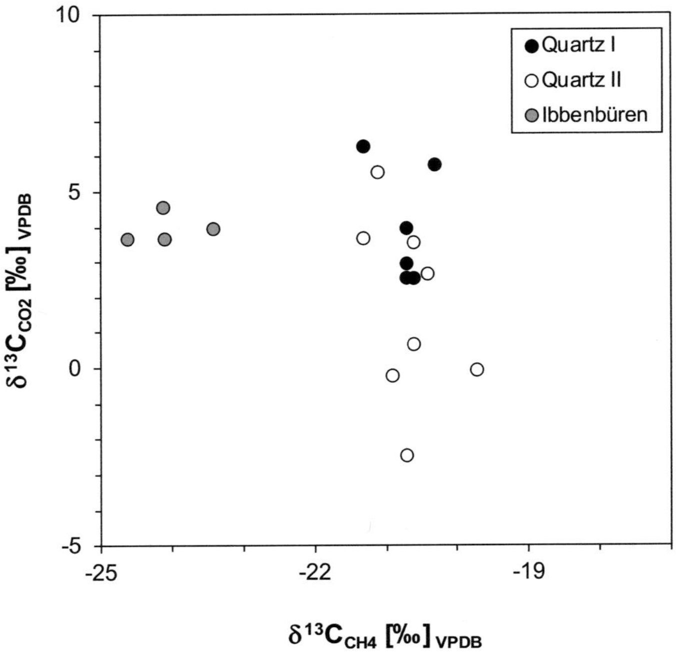

Figure 3.10: δ^{13}C values of methane (CH_4) plotted against δ^{13}C values of carbon dioxide (CO_2) measured in fluid inclusions hosted in quartz veins of the study area (quartz I and II) as well as of the Ibbenbüren coal district. The less negative δ^{13}C isotopy of CH_4 in the study area is maturity-related (c.f. Lüders et al., 2012).

3.5.5 K-Ar ages of illite

K-Ar illite age determination returned 161.8 Ma on average (ranging from 157.2 to 165.0 Ma, Tab. 3.1) for the < 2 μm fraction of all samples. The < 0.2 μm fraction of samples PB 29 and PB 53 yielded ages of 106.7 and 96.5 Ma, respectively (Tab. 3.1). The age determination on both size fractions was carried out in order to distinguish different illite generations (e.g. Wilkinson and Haszeldine, 2002).

Table 3.1: K-Ar data of the studied sandstone size fractions.

Sample	Fraction [µm]	K_2O [Wt. %]	^{40}Ar [nl/g] STP	^{40}Ar [%]	Age [Ma]	2σ-Error [Ma]	2σ-Error [%]
PB 14	< 2	4.38	24.39	96.52	165.0	4.1	2.5
PB 29	< 2	3.38	17.91	85.17	157.2	4.2	2.7
PB 29	< 0.2	8.65	30.65	93.9	106.7	8.7	8.1
PB 53	< 2	4.27	23.69	92.71	164.4	3.6	2.2
PB 53	< 0.2	9.96	31.85	96.14	96.5	1.0	1.0
PV 8	< 2	5.89	31.92	93.51	160.6	3.8	2.4

3.5.6 Vitrinite reflectance

Vitrinite measurements (Tab. 3.2) were conducted on samples taken at different positions throughout the quarry. Being more reliable and to account for the high maturity of the samples with concurrent increased vitrinite reflectance anisotropy, the rotational reflectance (VR_{rot}) was evaluated together with maximum and minimum reflectance (VR_{max}, VR_{min}, respectively) (e.g. Houseknecht et al., 1993; Houseknecht and Weesner, 1997; Littke et al., 2012). VR_{max} is a better indicator of maximum temperatures as it is less affected by tectonic stresses (Kilby, 1988). Both VR_{rot} (4.2 – 4.9) as well as VR_{max} (6.6 – 6.9) are in the same range for all measured samples with a maximum VR_{max} of up to 8.0 % (Tab. 3.2). The vitrinite data point to highest maturity of the investigated coals, reaching the meta-anthracite stage (e.g. Teichmüller et al., 1984). Furthermore, the similar ranges indicate a homogeneous peak temperature distribution throughout the exposed quarry. VR_{max} and VR_{min} reveal a high reflectance anisotropy (VR_{bi} = 4.7) for the sample set (Fig. 3.11, Tab. 3.2), typical for highly mature coals (e.g. Houseknecht and Weesner, 1997). The reflectance-indicating surface (RIS) cross-plots (Fig. 3.11), after Kilby (1988), provide evidence of a biaxial vitrinite bireflectance and the numerically derived parameters, RIS style (R_{st}, -1.0 – 0.8) and RIS amplitude (R_{am}, 0.2 - 0.3), point to a biaxial (even) vitrinite signature (cf. Kilby, 1988; Levine and Davis, 1989).

Table 3.2: Studied sample set of coals comprising reflectance readings based on 100 measurements per sample. [VR_{max} = average maximum vitrinite reflectance; VR_{min} = average minimum vitrinite reflectance; VR_{rot} = rotational vitrinite reflectance; VR_{bi} = bireflectance ($VR_{max} - VR_{min}$); max VR_{max} = highest measured vitrinite reflectance value; min VR_{min} = smallest measured vitrinite reflectance values; max VR_{bi} = max VR_{max} – min VR_{min}; Rst = RIS (reflectance-indicating surface) style; Ram = RIS amplitude]

Sample	VR_{max}	σ	VR_{min}	σ	VR_{rot}	VR_{bi}	max VR_{max}	min VR_{min}	max VR_{bi}	R_{st}	R_{am}
	[%]	[VR_{max}]	[%]	[VR_{min}]	[%]	[%]	[%]	[%]	[%]		
PB K1	6.57	0.36	2.33	0.45	4.15	4.24	7.48	1.54	5.94	- 1.00	0.22
PB K2	6.86	0.44	2.07	0.47	4.82	4.79	7.72	1.17	6.55	- 0.96	0.23
PB K3	6.89	0.21	2.11	0.25	4.88	4.78	7.50	1.42	6.08	- 0.44	0.23
PB K4	6.81	0.25	1.99	0.24	4.80	4.82	7.95	0.93	7.02	0.85	0.27
Average	6.78	0.32	2.13	0.35	4.66	4.66	7.66	1.27	6.40	- 0.39	0.24
σ	0.12	0.09	0.13	0.11	0.30	0.24	0.19	0.24	0.42	0.75	0.02

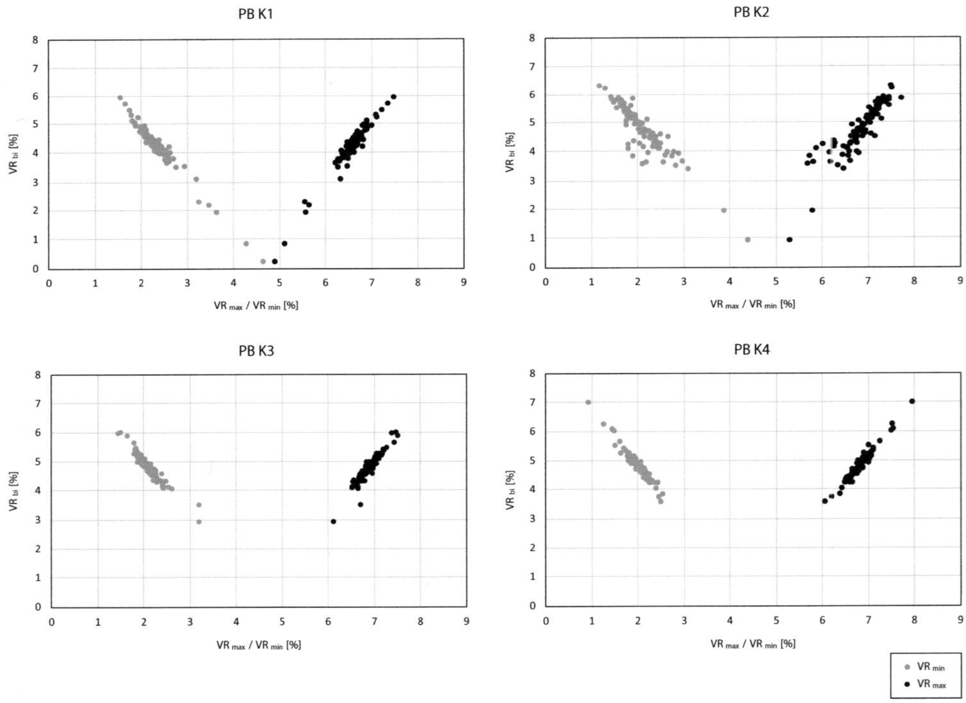

Figure 3.11: RIS plots of VR_{max} and VR_{min} versus VR_{bi} for the measured coal samples.

3.6 Discussion

3.6.1 Characterization of thermal events

Mineral assemblages, such as quartz cementation of illite meshes (Fig. 3.4c) and sericite, throughout the study area as well as chlorite imply higher temperatures reaching anchimetamorphic conditions (high-grade diagenesis and earliest metamorphism). This is further supported by chlorite thermometry (> 300°C for vein chlorite, ca. 270°C for host rock chlorite; Fig. 3.7) as well as vitrinite reflectance data (VR_{rot} ~ 4.7%, Tab. 3.2). Chamosite (Fig. 3.6) is considered to have formed under hydrothermal conditions or low-temperature metamorphism within iron-rich rocks (Deer et al., 1992; Spoetl et al., 1994). The presence of the IIb polytype can be interpreted as high-temperature modification of this chlorite species (Spoetl et al., 1994). The lower chlorite model temperatures in the host rock compared to the veins may indicate that transport of hot fluids was focused by fracture-flow. Chlorite and vitrinite geothermometer assessments are largely in accordance with trapping model temperatures (assuming fluid immiscibility) of aqueous two-phase inclusions in quartz I, which indicate fluid entrapment temperatures of about 250°C (or even higher if assuming fluid miscibility) and pressures of approximately 1 kbar (Fig. 3.12). Considering lithostatic pressure conditions of about 2.3 g/cm^3, which is likely since the Upper Carboniferous strata was covered by impermeable Zechstein cap rocks, precipitation of quartz I would have occurred at approximately 4.4 km depth (Fig. 3.13). Since pore-filling chlorite yields crystallization temperatures of about 270°C, a thermal gradient of approximately 61°C/km can be derived.

Upper Carboniferous sediments in the nearby Woitzel quarry 15 km away (Fig. 3.1) did not experience temperatures exceeding 180°C as indicated by lower vitrinite reflectance data between 1.6 and 1.9% R_m for the Westphalian C and D and the presence of coals of the semi- to anthracite stage (cf. Teichmüller, 1986). The lower matrix porosity and permeability in the Piesberg quarry compared to the Woitzel quarry (Fig. 3.5) may be thus attributed to the higher maximum temperature overprint, which is indicated by both the authigenic mineral inventory as well as the intensity of the resulting diagenetic modifications.

The fluid inclusion analyses and isochoric projections of quartz II (Fig. 3.12) indicate different pressure and temperature (P-T) conditions for fluid entrapment and point to a second event of vein formation with model entrapment temperatures of about 180°C and pressures of ca. 400 bar. The pressure variation might indicate a fluctuation of the pressure regime from lithostatic to hydrostatic (Fig. 3.13) and was most likely related to stages of basin inversion (Lüders et al., 2012). This is further evidenced by the not entirely sealed quartz II veins, which are partly inter-grown with ore minerals, such as marcasite.

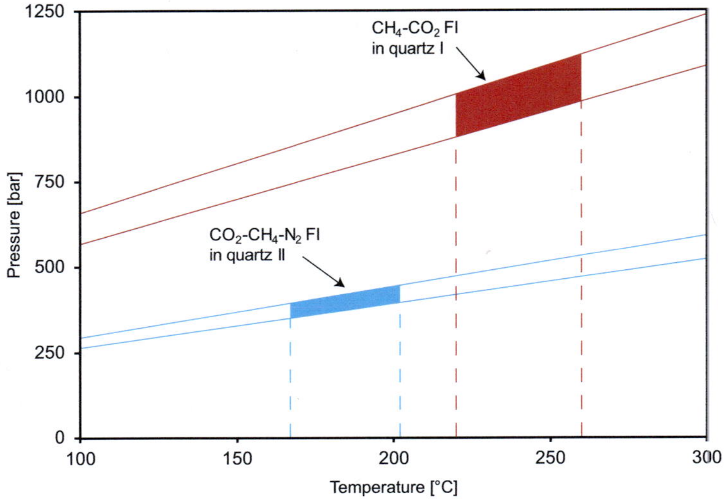

Figure 3.12: P-T diagram with isochoric reconstruction for the estimation of trapping conditions of co-genetically trapped primary aqueous two-phase (dotted lines) and gaseous (solid lines) fluid inclusions hosted in quartz I (red) and quartz II (blue). Assuming fluid immiscibility, the homogenization temperatures of aqueous two-phase inclusions are also the trapping temperatures (Roedder, 1984). In the case of fluid miscibility, the trapping temperatures would be even higher.

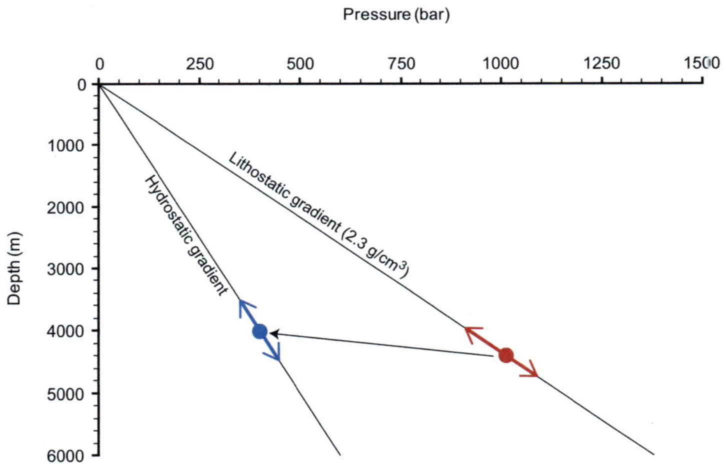

Figure 3.13: Projection of possible fluid entrapment conditions for gaseous and co-genetically trapped gas-bearing aqueous fluid inclusions hosted in quartz I (red) and quartz II (blue) derived from Fig. 3.12. Aqueous two-phase fluid inclusions in quartz II were possibly trapped at hydrostatic pressure during basin inversion as indicated by the presence of low-density C-H-O-N gas inclusions. Note that the formation depths of both quartz generations are similar, which suggests that Zechstein cap rocks had become permeable during basin inversion.

3.6.2 Timing of thermal events

The studied samples for K-Ar age determinations (Tab. 3.1) likely represent a mixture of detrital and authigenic mineral phases. A partial loss of radiogenic ^{40}Ar may occur when illite experiences temperatures exceeding 200°C (Odin and Bonhomme, 1982). Hunziker et al. (1986) stated that temperatures of 260 ± 30°C may even lead to the total reset of the K-Ar system. At temperatures of 300°C or higher, evidenced by chlorite geothermometry, it is likely that the derived Middle to Late Jurassic ages of the < 2 μm fraction might represent and depict the timing of a first thermal event (quartz I/chlorite in veins). Based on the higher nominal closure temperatures of muscovite of 350°C (Purdy and Jäger, 1976; Jäger, 1979), or even 425°C according to Harrison et al. (2009), the contribution of detrital components, such as metamorphic muscovite, cannot be excluded and might cause older K-Ar ages. However, their presence in this size fraction and thus a significant contribution to the measured K-Ar ages is very unlikely. The Middle to Late Jurassic ages of illite growth are in accordance with K-Ar dating of nearby Upper Carboniferous and Rotliegend reservoirs, which postulates Early Jurassic continuous illite formation between 210 and 180 Ma and Middle Jurassic (160 Ma) illite formation (Gaupp et al., 1993; Zwingmann et al., 1998; Liewig and Clauer, 2000).

The younger Early/Late Cretaceous K-Ar ages of the < 0.2 μm fraction (Tab. 3.1) may denote the crystallization of a later illite generation. This may have been caused by the growth of younger fibrous illite over older one, separately building up new nuclei. Thus, the illite did not form by continuous growth (e.g. Wilkinson and Haszeldine, 2002). The younger illite ages may represent the timing of the second thermal event (quartz II, T ~ 180°C) and additionally the youngest age for authigenic illite growth.

3.6.3 Fluid source and heat flow mechanism

The $\delta^{13}C_{(CO2)}$ values of fluid inclusion gases in both quartz generations strongly suggest an inorganic origin of CO_2 (Fig. 3.10) (Dai et al., 1996; Krooss et al., 2008). The slightly negative to positive values indicate decarbonation of marine carbonates or a metamorphic source for the CO_2 (Stahl, 1971; Dai et al., 1996; Hoefs, 2009; Duschl et al., 2016). In any case, both scenarios point to a deep-seated source of inorganic CO_2. The assumption of decarbonation or dissolution (e.g. Krooss et al., 2008) of (marine) carbonates can be well-explained by ascending fluids originating from or flowing through compact limestone of Early Carboniferous or Devonian age (Fig. 3.3) (e.g. Drozdzewski and Ribbert, 2003). During metamorphic decarbonation reactions at temperatures well above 300°C, strong carbon isotope fractionation occurs between the source carbonates and the released gaseous CO_2 phase (Bottinga, 1969). The latter becomes enriched in ^{13}C, resulting in an increase in $\delta^{13}C_{(CO2)}$ values by approximately 2.5 to 3‰ compared to the parental carbonate source and would explain the often measured positive $\delta^{13}C_{(CO2)}$ values of fluid inclusion gases in this work.

Based on magnetotelluric measurements in the North German Basin, Hoffmann et al. (2005, 2008) concluded the existence of Middle to Upper Devonian as well as Lower Carboniferous carbonates, which might form part of a "Lower Saxony Carbonate platform" (Hoffmann et al., 2005). Refraction seismic data (Brockamp, 1967) of the region of Bramsche (north of Osnabrück) first supported the hypothesis of a plutonic intrusion referred to as "Bramsche Massif" (e.g. Bartenstein et al., 1971; Stadler and Teichmüller, 1971) being responsible for noteworthy gravimetric and magnetic anomalies in the region (e.g. von Flotow et al., 1931; Reich, 1933; Breyer, 1971; Bachmann and Grosse, 1989). However, the ratio between compressive and shear waves deduced from refraction seismic (v_p/v_s = 1.72) (cf. Brockamp, 1967; Brink, 2013) rather points to dolomite (v_p/v_s = 1.6 – 1.75) than to a gabbroic intrusion (v_p/v_s = 1.85 – 1.95) (cf. Angenheister, 1982; Brink, 2013). This supports the presence of a Paleozoic carbonate platform (e.g. Hoffmann et al., 2005; Hoffmann et al., 2008; Brink, 2013).

The ascent of fluids and heat supply was most likely controlled by large-scale, deep-reaching faults. Lohr et al. (2007) described dextral transtension along NW-SE striking faults during the Middle/Upper Jurassic in the region. This timing, as well as the fault strike, are in line with the first thermal K-Ar event at 160 Ma and the presence of the major NNW-SSE striking eastern fault in the study area (Fig. 3.2). This fault, with a displacement of up to 600 m (e.g. Hinze, 1979; Baldschuhn et al., 2001), may thus have acted as a conduit for the heat/fluid source of the thermal overprint, associated with the precipitation of quartz I and chlorite. The presence of chamosite additionally bolsters the model of a hydrothermal heat source (Deer et al., 1992; Spoetl et al., 1994).

The CO_2-rich fluid inclusions hosted in quartz II fracture mineralization (Fig. 3.9), intergrown with ore minerals, like marcasite, point to a second event of ascending iron- and CO_2-rich fluids through the tight gas reservoir analog. This event might be associated with the K-Ar age of the < 0.2 μm fraction (Tab. 3.1) and with WSW-ESE striking faults. In a graben structure with this fault strike, Zechstein carbonates were altered to siderite with brown iron ore (Lotze, 1953; Voigt, 1960).

3.6.4 Deep burial versus hydrothermal fluid flow

The high-temperature overprint of coaly material in the study area has been related to deep burial of locally more than 7 km with high heat flow from the Late Jurassic until the Early Cretaceous and subsequent basin inversion and uplift (e.g. Petmecky et al., 1999; Senglaub et al., 2006; Bruns et al., 2013). The maximum temperature of 180°C (see Chapter 3.6.1) of the Woitzel quarry 15 km to the west in the Ibbenbüren hard coal mining district corresponds to a 4 – 5 km burial. Considering the same stratigraphic age and proximity of the adjacent Woitzel quarry, the different burial models would require a deeper burial of the study area from the Triassic to Cretaceous and an uplift of the same amount since the Late Cretaceous. Evidences for the associated thicker Mesozoic syn-rift overburden and increased sediment erosion since the Upper Cretaceous around the study area are lacking. Furthermore, the significant lower maximum temperatures in the adjacent quarry would imply a high lateral temperature gradient. Therefore, a burial-related maturation model of organic matter seems improbable. Instead, a local temperature phenomenon, also observed in other parts of the Lower Saxony Basin (Reutel et al., 1995; Will et al., 2016), may have brought about the high-temperature anomaly. Taking into account the homogeneous vitrinite reflectance across the quarry (see Chapter 0), the lateral thermal overprint is on a kilometer-scale away from the fault (approximately 1 km in the W-E direction, Fig. 3.2) and thus a reservoir-sized anomaly. This is also well-documented by diagenetic features, which can be likewise observed throughout the quarry, such as the intensely cemented fibrous illite meshwork (Fig. 3.4c).

The $\delta^{13}C$ (CH_4) values of gaseous inclusions in quartz from the Ibbenbüren coal mine (Fig. 3.10) are typical for thermogenic methane from Carboniferous coal measures of many gas reservoirs of the Lower Saxony Basin (e.g. Stahl, 1977). The less negative $\delta^{13}C$ (CH_4) values of gaseous inclusions in both quartz generations from the study site (Fig. 3.10) suggest that the difference in the $\delta^{13}C$ (CH_4) values from the Ibbenbüren coal mine and Piesberg is maturity-related (Lüders et al., 2012). In the study area, however, thermogenic methane was derived from coal measures that experienced highest maturity ($VR_{rot} \sim 4.7$ %, Tab. 3.2) at the time of precipitation of both quartz generations. Thus, the $\delta^{13}C$ (CH_4) values of the fluid inclusion gases in quartz I (Fig. 3.10) point to a maximum temperature and maturation of Carboniferous coals at the Piesberg during a fluid event that also may have caused crystallization of authigenic illite and chlorite in Mid/Late Jurassic times, i.e. prior to peak subsidence (Tab. 3.1) (q.v. Gaupp and Okkerman, 2011). Subsidence models based on equilibrium temperatures relate the high maturity of the anthracite coals to maximum burial and heat flow in the early Late Cretaceous, which implies a subsequent erosion of approximately 6700 m basin infill during uplift after Late Cretaceous basin inversion (e.g. Bruns et al., 2013). Conversely, a fault-related temperature increase, with maximum temperatures of about 300°C or more, requires a burial depth of about 4400 m assuming a lithostatic pressure regime.

A deep burial model with highest coalification at peak subsidence in the Early Cretaceous implies an extensional regime without any significant additional compressive tectonic forces (c.f. Kley and Voigt, 2008). If coalification were caused by such a “pure” deep burial, the coals would feature an uniaxial (negative) reflectance anisotropy (e.g. Levine and Davis, 1984; Kilby, 1988; Levine and Davis, 1989; Houseknecht and Weesner, 1997).

Contrarily, the rotational vitrinite reflectance measurements and numerically derived RIS style (R_{st}) and reflectance anisotropy magnitude (R_{am}), after Kilby (1988), point to a biaxial (even) vitrinite reflectance anisotropy (Tab. 3.2, Fig. 3.11). A transtensional regime as postulated for the Mid- to Late Jurassic (e.g. Lohr et al., 2007) better explains the observed coalification pattern. Similar results were presented by Frings et al. (2004) on Upper Carboniferous coal measures in NW Spain. Overall, the presented data as well as $\delta^{13}C$ (CH_4) values in the fluid inclusions hosted within both quartz I and II indicate that present day maturity was likely already attained with the precipitation of quartz I (~250°C) in the Mid- to Late Jurassic and that the anthracite coal maturity was most likely not significantly modified until peak subsidence.

3.6.5 Fault-bound hydrothermal fluid flow

Assuming a fault-bound hydrothermal overprint, the ascending fluid flow and heat transfer is most likely restricted to the fault damage zone of the large-scale, 600 m displacement eastern fault, where associated fractures acted as fluid conduit (Fig. 3.14) (e.g. Odling et al., 2004; Faulkner et al., 2010). Torabi and Berg (2011) summarized the scaling relationships of fault attributes (e.g. Cowie and Scholz, 1992; Fossen and Hesthammer, 1997; Bonnet et al., 2001; Schultz and Fossen, 2002; Odling et al., 2004; Shipton et al., 2006; Faulkner et al., 2010, 2011). Published data suggest that fault damage zones of such large-scale faults are several tens to hundreds of meters wide (e.g. Shipton et al., 2006; Torabi and Berg, 2011).

A displacement of 600 m would equal a horizontal fault length of approximately 10 km following the scaling relationship $D = 0.03L^{1.06}$ (D = displacement, L = horizontal fault length, perpendicular to dip) as proposed by Schlische et al. (1996). This range of fault length is also visible from maps of the Lower Saxony Basin (Fig. 3.1) (cf. Baldschuhn and Kockel, 1999). An estimated length-to-height ratio of 2 – 3 (e.g. Nicol et al., 1996; Soliva et al., 2005; Schultz et al., 2006) indicates a fault height from the point of maximum displacement (of at least 600 m) of approximately 2.5 km. Assuming that the normal geothermal gradient in the region during the Jurassic (e.g. 30°C/km, Woitzel quarry) was deflected in the study area, increasing temperatures by 90 - 120°C, as indicated by the anomalous high chlorite geothermometers of ca. 300°C, the fault zone must have been connected to a source some 3 km deeper. This is in agreement with the estimated fault height determined herein.

An advection-driven vertical fluid flow and heat transfer requires a bulk transport of mass or heat and thus a reservoir/source. Contrarily, a convective model would solely rely on the buoyancy contrast in a fault-bound fracture zone. Convection-driven models in fault zones were already discussed for natural examples and theoretical studies (e.g. Bjørlykke et al., 1988; Fleming et al., 1998; Simms and Garven, 2004; Timar-Geng et al., 2009; Li et al., 2016) and avoid large fluid volume recharge required by advection driven fluid flow.

The increased temperatures across the Piesberg quarry over a minimum lateral distance of 1 km may be the result of lateral heating originating from the fault damage zone (Fig. 3.14). Considering an average thermal diffusivity of sandstones of 0.8 mm^2s^{-1} at 300°C (cf. Hartlieb et al., 2016), a time span of approximately 40,000 years may be estimated for the pure conductive propagation of the temperature change over a distance of 1 km based on the simplified relationship $\tau = l^2/K$ (τ = time, l = distance, K = thermal diffusivity) (cf. Turcotte and Schubert, 2002). This reasonable geologic time frame may well explain lateral heating of the sandstones as a result of increased temperatures because of fluid circulation in the fault damage zone of the large-scale eastern fault (Fig. 3.14).

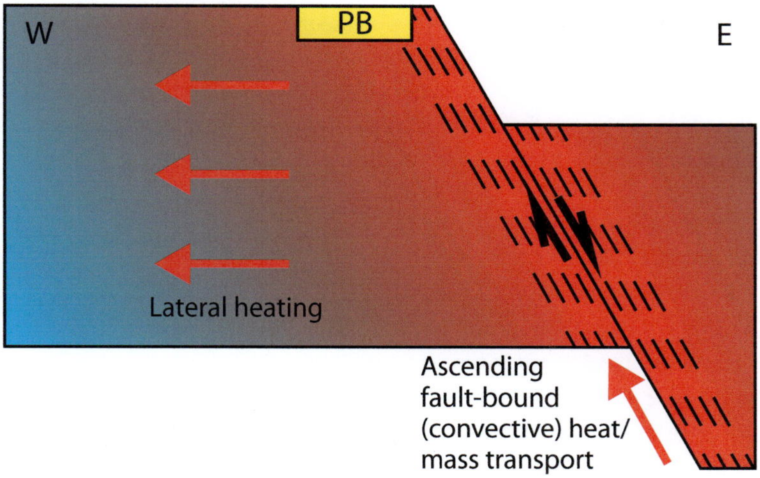

Figure 3.14: Sketch (not to scale) illustrating the fault-bound thermal anomaly in the study site. The fault damage zone (dashed lines) of the large NNW-SSE striking fault is several tens to hundreds of meters wide and may provide enhanced fracture permeability for the (convective) heat/mass transport. In a consequence of the increased temperature in the fault damage zone, the rocks were laterally heated up. The study site is represented by the yellow rectangle (PB) and it is noted that at least the entire outcrop was affected by the thermal overprint, implying at least a kilometer-scale thermal anomaly.

3.6.6 Implications for petroleum system modeling and exploration

Numerical basin modeling in sedimentary rocks is generally based on temperature indicators, such as vitrinite reflectance, and therefore frequently relies on available core material (e.g. Welte and Yalçin, 1988; Yalçin et al., 1997; Littke et al., 2000; Allen and Allen, 2005; Harris and Peters, 2012). Assuming a closed system in equilibrium, the maturity of organic matter is linked to the overburden and estimated geothermal gradient, from which the subsidence curve can be reconstructed (e.g. Welte and Yalçin, 1988; Yalçin et al., 1997; Littke et al., 2000). The model provides a burial scenario and thermal reconstruction over time, which are required to predict the estimated oil-gas resources (e.g. Allen and Allen, 2005; Hantschel and Kauerauf, 2009; Harris and Peters, 2012; Bruns et al., 2013). The diagenetic, thermal and age data imply that highest temperatures were not associated with the deepest burial in the studied tight gas siliciclastics (q.v. Gaupp and Okkerman, 2011). Thermal anomalies as result of fault-bound circulating hot fluids must be taken into account for the modeling of petroleum systems. Assumed steady-state temperatures based on conductive heat transfer are appropriate for many scenarios and are thus a reasonable best practice modeling approach for many petroleum systems.

However, thermal anomalies around large fault systems, such as present in the study site, can laterally reach a kilometer scale and may be a reservoir risk. Observations of thermal anomalies are well-known in the Lower Saxony Basin (e.g. Teichmüller, 1986; Sedat, 1992; Will et al., 2016) as well as in Upper Carboniferous tight gas fields (e.g. Reutel et al., 1995). However, the lateral extent has remained unknown and the fault-related origin has been hardly discussed so far (e.g. Senglaub et al., 2005, 2006; Bruns et al., 2013). Since thermal anomalies are also reported in other sedimentary basins worldwide (e.g. Andrews et al., 1996; Fleming et al., 1998; Frings et al., 2004), especially in extensional settings such as the Basin and Range region (United States) (Wisian and Blackwell, 2004) or the European Cenozoic rift system (Lampe and Person, 2000; Timar-Geng et al., 2004, 2009; Lampe and Schwark, 2012), these exceptional heat sources and their impact on reservoir quality must be considered in petroleum modeling approaches. Timar-Geng et al. (2009) demonstrated that convective heat transfer by fluid flow has a large impact and may result in significant overestimations of erosion rates of up to 80%. Therefore, in the presented case study, reservoir modeling with the highest temperatures at peak subsidence in Early Cretaceous times based on calibration with vitrinite reflection data may overestimate the true burial depth.

3.7 Conclusions

The tight gas siliciclastics show indications of a high-temperature thermal overprint of about 250 – 300°C based on chlorite thermometry, fluid inclusion data and vitrinite reflectance measurements around a large NNW-SSE striking normal fault with 600 m displacement. The local thermal increase of approximately 90 – 120°C was the result of hydrothermal fluids circulating along the fault damage zone of this large fault system, which laterally heated up the tight gas sandstones. Vitrinite reflectance data of meta-anthracite coals highlight the lateral extent of the high-temperature anomaly of up to a one-kilometer distance from the large fault. This is further yielded by the temperature-related diagenetic overprint of the entire study site, highlighting the exploration risk of such thermal anomalies around faults and demonstrating the deterioration of reservoir properties with respect to porosity and permeability.

Highest coalification was reached prior to peak burial during Mid to Late Jurassic rifting (162 Ma, K-Ar age determination on < 2µm fraction) and coalification was not subject to significant changes during the ongoing burial history as reflected by stable isotope data from fluid inclusions and biaxial reflectance anisotropy of meta-anthracite coals. The precipitation of a first generation of quartz close to lithostatic fluid pressure with temperatures of approximately 250°C together with authigenic chlorite growth in mineralized extension fractures was most likely contemporaneous with authigenic illite and chlorite growth in the host rock. Quartz cementation of illite meshes and K-Ar ages of the < 0.2 µm sample fraction suggest ongoing burial until the Late Cretaceous (90 – 100 Ma). A contemporaneous second quartz generation precipitated in extension fractures under hydrostatic fluid pressure conditions and temperatures of about 180°C.

Our study demonstrates that peak temperatures do not necessarily imply peak subsidence and consequently that fault-related local thermal anomalies need to be incorporated into the modeling of petroleum systems and reservoir quality prediction modeling of tight gas sandstones. Ignoring fault-related phenomena affecting reservoir properties laterally over more than one kilometer may overestimate recoverable resources.

4 Evaluation of a workflow to derive t-LiDAR fracture statistics of a tight gas sandstone reservoir analog

4.1 Abstract

Understanding natural fracture networks in the subsurface is highly challenging, as direct 1D borehole data are unable to reflect their spatial complexity, and 3D seismic data are limited in spatial resolution to resolve individual meter-scale fractures.

Here, we present a prototype workflow for automated fracture detection along horizontal scanlines using terrestrial laser scanning (t-LiDAR). Data are derived from a kilometer-scale Upper Carboniferous reservoir outcrop analog in the Lower Saxony Basin, NW Germany. The workflow allows the t-LiDAR data to be integrated into conventional reservoir modeling software for characterizing natural fracture networks with regard to orientation and spatial distribution. The analysis outlines the lateral re-orientation of fractures from a WSW-ENE strike, near a normal fault with approximately 600 m displacement, towards a W-E strike away from the fault. Fracture corridors, 10 to 20 m wide, are present in unfaulted rocks with an average fracture density of 3.4 – 3.9 m^{-1}. A reservoir-scale digital outcrop model was constructed as a basis for data integration. The fracture detection and analysis serve as input for a stochastically-modeled discrete fracture network (DFN), demonstrating the transferability of the derived data into standard hydrocarbon exploration and production industry approaches.

The presented t-LiDAR workflow provides a powerful tool for quantitative spatial analysis of outcrop analogs, in terms of natural fracture network characterization, and enriches classical outcrop investigation techniques. This study may contribute to a better application of outcrop analog data to naturally fractured reservoirs in the subsurface, reducing uncertainties in the characterization of this reservoir type at depth.

4.2 Introduction

The Central European Basin System hosts a large number of Paleozoic gas fields, reaching from the North Sea through the Netherlands and Germany to Poland (Glennie, 1986, 1997;

Gaupp et al., 2008; Littke et al., 2008, 2011). Low-permeability reservoirs, such as the tight gas sandstones of the Upper Carboniferous, have become more common for gas exploration in the last few decades (Leeder and Hardman, 1990; Besly, 1998; Petmecky et al., 1999; Kombrink et al., 2010). So-called 'sweet-spot' prediction in these reservoirs is a challenge because the reservoir characteristics of these formations are still not well understood (Cowan, 1989; Besly, 1998). With typically low permeability, careful characterization of the natural fracture network is indispensable for estimating flow properties and, thus, successful exploitation (e.g. Bourbiaux, 2010; Bahrami et al., 2012; Santos et al., 2015). A challenge is to characterize the sub-seismic-scale structural inventory, which remains a major issue for reservoir modeling (Seers and Hodgetts, 2013).

Analog studies overcome the limits in the resolution of seismic sections and the information from core material (e.g. Miall, 1988a; Sharp et al., 2003; Pranter et al., 2007, 2014). This applies even more to the structural characterization as structural heterogeneities are often below seismic resolution (e.g. Hodgetts, 2013; Seers and Hodgetts, 2013). The incorporation of analog studies has tremendously enhanced the understanding of the structural style and heterogeneity of their subsurface reservoir counterparts (e.g. Köstler and Ehrmann, 1991; Sharp et al., 2003; Olson et al., 2009). Outcrop data have been used to unravel spatial characterization concerning the geometry and distribution of natural fracture networks, which cannot be directly assessed in the subsurface (e.g. Odling et al., 1999; Florez-Nino et al., 2005; Ghosh and Mitra, 2009; Gillespie et al., 2011). The conventional manual analysis of fracture networks, based on fieldwork in outcrop analogs, is time-consuming and workforce-intensive and suffers often from the inaccessibility of sampling areas (e.g. Priest and Hudson, 1981; Marrett et al., 1999; Ortega and Marrett, 2000; McCaffrey et al., 2005; Ortega et al., 2006; Guerriero et al., 2011; Seers and Hodgetts, 2013; Becker et al., 2014; Casini et al., 2016).

The first efforts to incorporate digital geology technology into the petroleum industry, presented by Bryant et al. (2000), dealt with digital outcrop models for comparison with subsurface datasets. Nowadays, the use of digital outcrop models, based for example on the technology of terrestrial laser scanning (t-LiDAR, Terrestrial Light Detection and Ranging) or photo-realistic models, is an established digital geology technology used in the geosciences for a broad range of studies (e.g. Xu et al., 2001; Bellian et al., 2005; McCaffrey et al., 2005; Olariu et al., 2005; Pringle et al., 2006, 2010; Enge et al., 2007, 2010; Labourdette and Jones, 2007; Redfern et al., 2007; Alfarhan et al., 2008; Buckley et al., 2008, 2010; Fabuel-Perez et al., 2009, 2010; Rotevatn et al., 2009; Wilson et al., 2009; Jones et al., 2011; Hodgetts, 2013; Becker et al., 2014; Rarity et al., 2014; Jacquemyn et al., 2015; Laux and Henk, 2015; Mahmud et al., 2015; Casini et al., 2016; Tavani et al., 2016). The Virtual Reality Geological Studio (VRGS) of David Hodgetts (University of Manchester, UK) is a dedicated software solution for working with digital outcrop models and respective data visualization and interpretation. Amongst others, it aims to integrate digital outcrop data with petroleum industry solutions (for case studies see Wilson et al., 2009; Fabuel-Perez et al., 2010; Seers and Hodgetts, 2013; Rarity et al., 2014). The application of digital outcrop models has resulted in substantial progress regarding quantitative geological mapping over large, inaccessible areas, and has enabled the analysis of a wide range of sampling scales (Fig. 4.1; e.g. Hodgetts, 2013; Rarity et al., 2014). Nevertheless, in most of the studies, the interpretation of virtual outcrops was conducted manually based on digital mapping techniques, for example (e.g. Fernández et al., 2004; Bellian et al., 2005; Wilson et al., 2006; Labourdette and Jones, 2007; Redfern et al., 2007; Fabuel-Perez et al., 2009, 2010; Rotevatn et al., 2009; Enge et al., 2010; Pringle et al., 2010; Jacquemyn et al., 2015).

T-LiDAR has been the most crucial technique for acquiring digital outcrops in the last decade (e.g. Pringle et al., 2006; Buckley et al., 2008; Jones et al., 2009; Cawood et al., 2017). It produces a vast database for natural fracture network characterization (e.g. Olariu et al., 2008; Koptíková et al., 2010; Lato et al., 2010; Gillespie et al., 2011; Wilson et al., 2011; Lato and Vöge, 2012; Becker et al., 2014; Laux and Henk, 2015; Casini et al., 2016) and can, amongst other things, provide fundamental parameters for the modeling of a DFN (e.g. Wilson et al., 2011; Hodgetts, 2013; Seers and Hodgetts, 2013; Laux and Henk, 2015; Casini et al., 2016). The manual interpretation of virtual outcrops is time-consuming work and, thus, does not exploit the entire provided potential, especially in large virtual outcrops (e.g. Casini et al., 2016). Therefore, considerable effort has been put into the development of (semi-)automated solutions for the recognition and description of fractures based on point cloud or mesh data (e.g. Kemeny and Donovan, 2005; Roncella and Forlani, 2005; Kemeny et al., 2006; van Knapen and Slob, 2006; Ferrero et al., 2008; Olariu et al., 2008; García-Sellés et al., 2011; Gigli and Casagli, 2011; Mah et al., 2011; Wilson et al., 2011; Vöge et al., 2013; Riquelme et al., 2014; Vasuki et al., 2014; Laux and Henk, 2015; Pless et al., 2015; Casini et al., 2016).

A major challenge is the direct extraction of fracture statistics/parameters from 3D point clouds. The first methods for the automated identification of fracture surfaces comprise the detection of connected triangles with the same orientation characteristics, based on triangulated meshes (e.g. Kemeny and Donovan, 2005) or the classification of point cloud data (e.g. Jaboyedoff et al., 2007). García-Sellés et al. (2011) applied the tensor analysis approach (Woodcock, 1977; Fernández, 2005) directly on point cloud data to identify and triangulate planar geological features with respective orientation data. The remaining difficulty is the direct derivation of fracture density parameters. A commonly applied approach to derive spatial data of natural fracture networks is the use of fracture trace maps, where fractures are digitized based on photographs (e.g. Barton et al., 1995; Le Garzic et al., 2011; Strijker et al., 2012; Hardebol and Bertotti, 2013; Bisdom et al., 2014). Several (semi-)automated approaches have been developed for fault and fracture trace mapping based on 3D datasets. These rely, for example, on ant tracking technology (e.g. Gillespie et al., 2011; Monsen et al., 2011), mesh-based methods (e.g. Umili et al., 2013) or approaches, which integrate orthorectified photos with digital elevation models (e.g. Vasuki et al., 2014). Several studies have combined such (previously listed) methods, for example different (semi-)automated digital or manual interpretation approaches, to improve resulting fracture models (e.g. Seers and Hodgetts, 2013, 2016a; Casini et al., 2016).

This work aimed to develop a hands-on and quick approach to unravelling rock anisotropy, concerning natural fractures for, and closely aligned to industry approaches and technology. We present a prototype workflow to derive fracture network statistics, based on a vast t-LiDAR dataset from a reservoir-scale outcrop analog for fluvial tight gas deposits of the region. This comprises an algorithm, which permits the automated detection of fracture discontinuity surfaces directly along virtual scanlines (sensu Voyat et al., 2006), based on acquired point cloud data. This allows for speedy processing and direct integration into modern reservoir modeling software and workflows. Besides the derivation of orientation measures, the presented approach provides spatial coordinates for each identified fracture. The data integration in a digital outcrop model, and the modeling of a DFN, demonstrates the applicability and practicability of the outlined workflow.

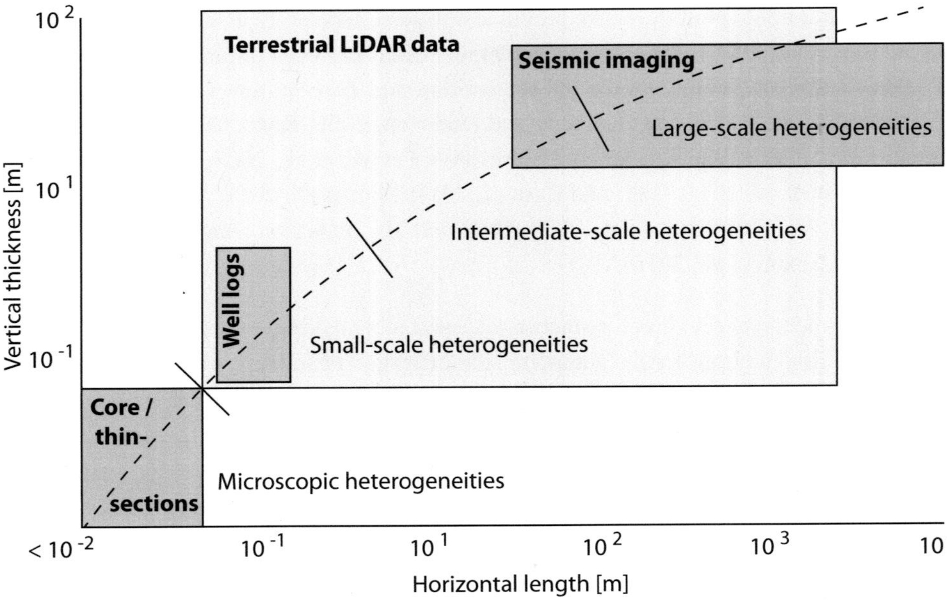

Figure 4.1: Different scales of heterogeneities in relation to sampling scales of seismic data, well logs and cores (after Rarity et al., 2014). Structural and sedimentological features can be attributed to a wide range of scales. T-LiDAR data facilitates the coverage of a wide scale range and can close a large gap between well data and seismic data (cf. Hodgetts, 2013).

4.3 Geological setting

4.3.1 The study area

Fracture data were investigated in the Piesberg Quarry near the town of Osnabrück, in the SW of the Lower Saxony Basin (Fig. 4.2). The study area belongs to the Late Variscan WNW-ESE striking Piesberg-Pyrmont axis (Drozdzewski, 1985; Drozdzewski et al., 2009). The Upper Carboniferous tight sandstones are separated from Triassic sediments in the east by a major NNW-SSE striking normal fault (Fig. 4.3) with a displacement of up to 600 m (Baldschuhn et al., 2001). Both fault orientations reflect the overall structural pattern of the region (Fig. 4.2) and locally show an early dip-slip and later strike-slip movement (e.g. Wüstefeld et al., 2017b). Gentle folding of the uplifted block was first explained by vertical movements (Hollmann et al., 1971) and more recently by dextral strike-slip movements on W-E directed faults (Fig. 4.3, Drozdzewski, 1998) as a result of Upper Cretaceous inversion tectonics, which is characteristic of the southern part of the Central European Basin System (Kley and Voigt, 2008; Sippel et al., 2009). For the study area, Drozdzewski (1998) and Bissen (2011) reported a main WNW-ESE fracture orientation as well as a NNW-SSE striking set for the entire quarry (see also Hollmann et al., 1971). In the eastern quarry, dominance of a WSW-ENE striking fracture set has been described (Drozdzewski, 1998; Bissen, 2011), while the WNW-ESE set is reported to be absent.

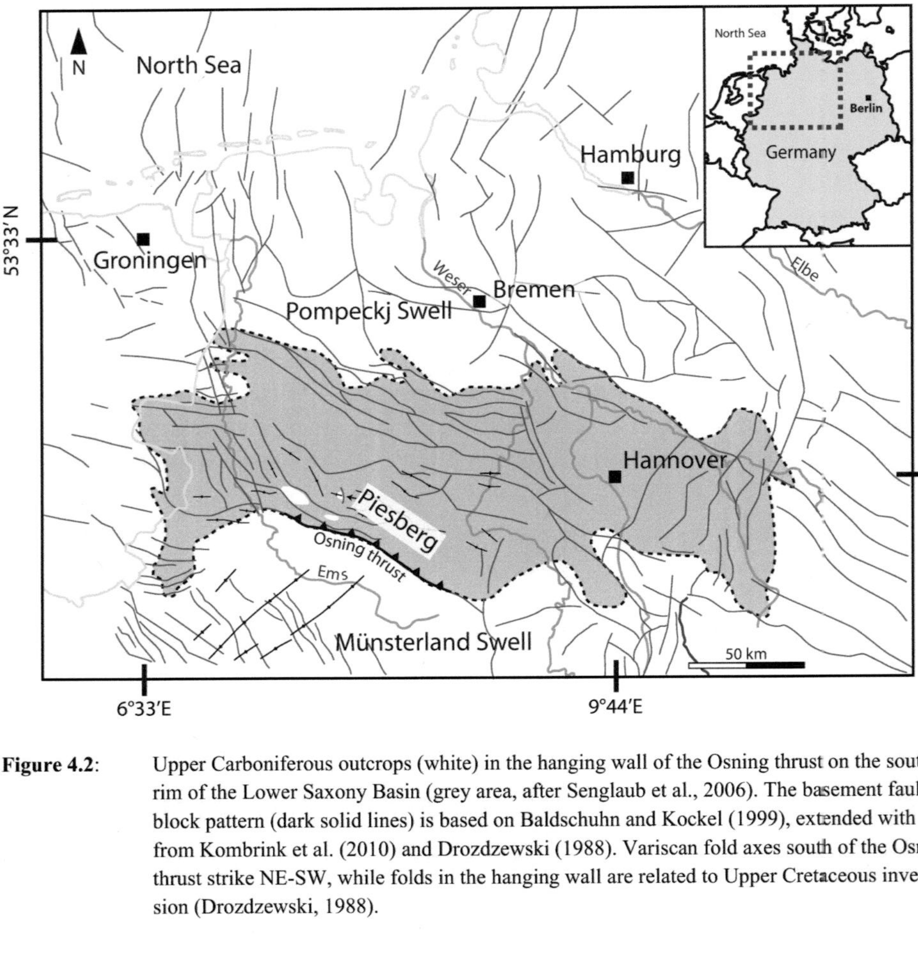

Figure 4.2: Upper Carboniferous outcrops (white) in the hanging wall of the Osning thrust on the southern rim of the Lower Saxony Basin (grey area, after Senglaub et al., 2006). The basement fault block pattern (dark solid lines) is based on Baldschuhn and Kockel (1999), extended with data from Kombrink et al. (2010) and Drozdzewski (1988). Variscan fold axes south of the Osning thrust strike NE-SW, while folds in the hanging wall are related to Upper Cretaceous inversion (Drozdzewski, 1988).

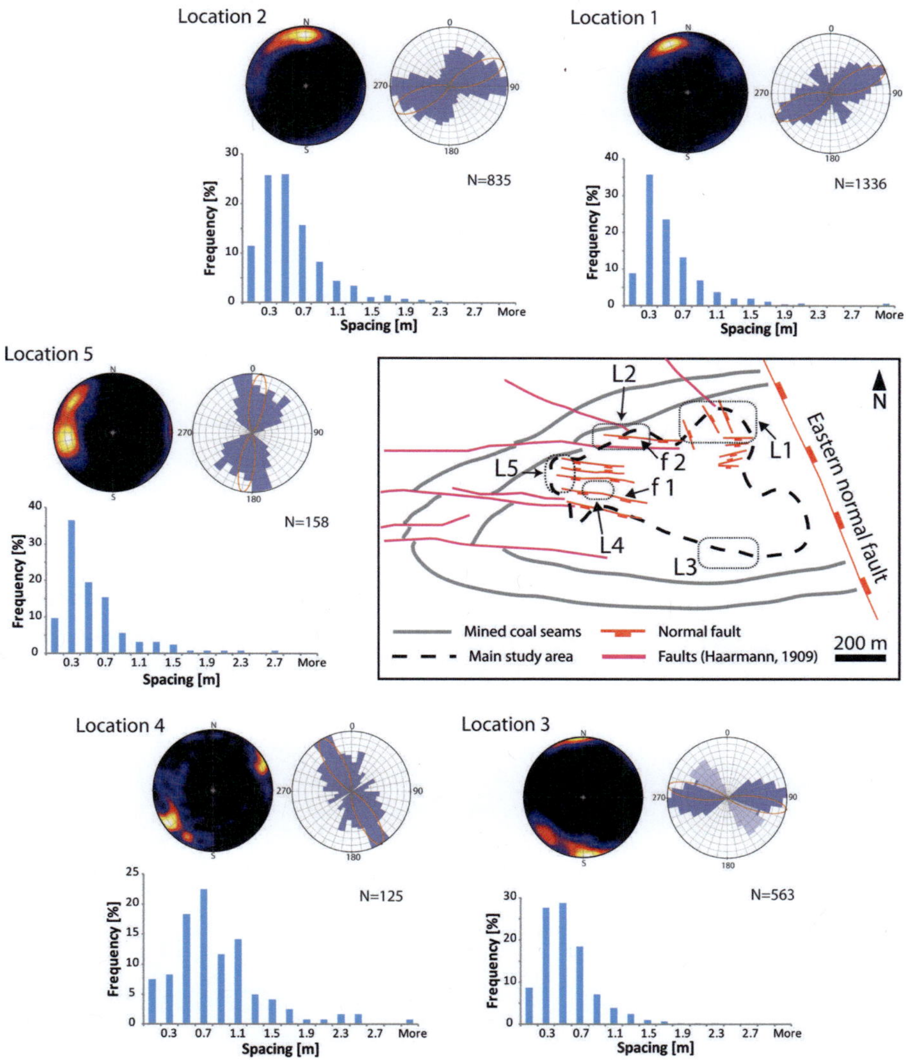

Figure 4.3: Simplified sketch of the quarry. Coal seams, faults, including the major normal fault in the east, are based on historical subsurface mining data (Haarmann, 1909). Normal faults with dip direction indicators were observed in this study (faults f1 and f2 are mentioned in the lateral fracture analysis). Locations L1 to L5 indicate the sampling sites for the virtual scanlines. Stereonet and strike diagrams, as well as fracture spacing frequency distributions, are provided for all locations.

4.3.2 Basin evolution

The basin evolution started in the Late Carboniferous (Pennsylvanian) with sedimentation in the foreland basin northwest of the Variscan orogeny (Ziegler, 1977; Glennie, 1986; Brink et al., 1992; Franke, 1995). The Upper Carboniferous Westphalian C and D exposed in the outcrop is dominated by cyclic fluvial deposits, comprising clastic sediments with intercalated coal seams (Fig. 4.4; David, 1987, 1990; Jones and Glover, 2005; Wüstefeld et al., 2014). Variscan compression resulted in transpressional structures in the Osnabrück/Ibbenbüren area as well as the hydrocarbon reservoir blocks Husum-Schneeren and Rehden (e.g. Drozdzewski et al., 2009). During the Permian, subsidence was mainly controlled by thermal relaxation of the lithosphere as well as sedimentation (Scheck and Bayer, 1999; van Wees et al., 2000; Bruns et al., 2013), and followed by rifting because of extension in the Triassic (e.g. Brink, 1992).

Inversion tectonics succeeded the deepest subsidence in the Late Cretaceous (Senglaub et al., 2005; Kley et al., 2008). This resulted in an uplift of the Upper Carboniferous, which led to exposure of the rocks at the surface in the hanging wall of the Osning thrust (Fig. 4.2), a system of normal faults that were reactivated to form thrust faults in the Late Cretaceous (e.g. Baldschuhn and Kockel, 1999; Kley et al., 2008). The stress field rotated during Upper Cretaceous basin inversion from a N-S/NE-SW to a NW-SE compressional direction (Kley and Voigt, 2008; Sippel et al., 2009). Today's stress orientation in the region is directed NNW-SSE (World Stress Map 2008, Heidbach et al., 2008).

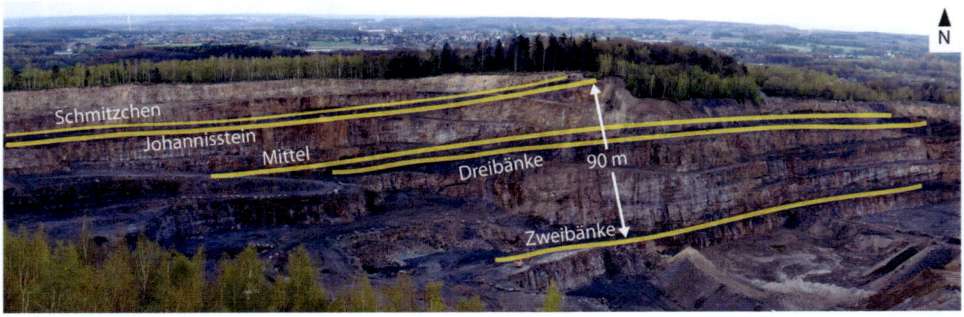

Figure 4.4: Northern quarry wall, exposing cyclic fluvial deposits of the Upper Carboniferous Westphalian C and D, which are separated by coal seams (after Wüstefeld et al., 2017a).

4.4 Methods

4.4.1 Outcrop data acquisition

T-LiDAR was employed to digitize the entire quarry to establish a geo-referenced 3D model (Fig. 4.5a). The scanner emits a laser pulse and measures the two-way travel-time for each reflection of the laser beam to create spatial coordinates for each measuring point, resulting in a 3D point cloud (Fig. 4.5; e.g. Bellian et al., 2005). We used a FARO Laser Scanner Focus 3D x330, with an operating range of up to 330 m, with a wavelength of the emitted laser pulse of 1550 nm and a sampling rate of up to 976,000 points per second. Two perpendicularly oriented axes of the scanner permit an operating range of 360° horizontally and -60° to 90° vertically. The resolution of the point cloud depends on the distance between the scanner and the measured surface and decreases with distance (e.g. Sturzenegger and Stead, 2009; Laux and Henk, 2015). The maximum angle between the two individual laser beams used in this study is 0.035° for individual scans, which results in a minimum resolution of 6.1 mm/10 m. This means that for each 10 m distance between laser scanner and quarry wall, the distance between two neighboring points increases by 6.1 mm. For example, 20 m distance equals a distance of 12.2 mm between adjacent points and a resolution of approximately 6700 points per square meter. In addition to the travel-time, the intensity of the reflected signal, relative to the emitted signal, was recorded (e.g. Pfeifer et al., 2008), and the resulting point cloud is represented in grayscale (Fig. 4.5c). Reflective properties rely, amongst other things, on the mineralogical composition of the individual sandstone grains, their weight fractions and grain size distributions (e.g. Clark and Roush, 1984; Bowitz and Ehling, 2008; Burton et al., 2011). Burton et al. (2011) demonstrated that the LiDAR intensity correlates, for example, with the quartz or clay content of investigated siliciclastic rocks. Coal seams are represented by relatively dark colors, i.e. low reflection intensities, whereas quartz-rich sandstones are characterized by bright gray colors (Fig. 4.5c). The scanner enables the simultaneous acquisition of photographs, thus providing a true coloring of the point cloud in addition to laser intensity measurements (Fig. 4.5c). Both signals represent valuable data for distinguishing different lithologies as well as weathering effects (Song et al., 2002; Burton et al., 2011). In order to digitize the entire quarry, the raw model consisted of 39 individual scans, which were chosen to avoid shading effects (e.g. Lato et al., 2009). The individual scans were registered and combined into a single point cloud (Fig. 4.5a) with the scanner software *FARO Scene*. Therefore, reference spheres were utilized and placed on the outcrop to act as common tie-points between scans (cf. Buckley et al., 2008; Hodgetts, 2013). The resulting point cloud, consisting of approx. 2×10^9 individual points, with a resolution of km- down to cm-scale, was georeferenced and quality-checked with differential GPS measurements (Topcon GR-5).

The precise northing was additionally guaranteed with reference objects, which were calibrated in the outcrop with a sighting compass (two sets of reference spheres, at approximately 50 m distance, which are accurately aligned N-S and W-E, respectively). In addition to the t-LiDAR-based analysis of the fracture network, fracture orientation and spacing were also measured traditionally with a geological compass.

4.4.2 Digital outcrop model

The acquired t-LiDAR point cloud was homogenized and reduced to a resolution of at least 1 cm (0.39 in) distance between individual points on average, using the open-source software *CloudCompare* (Girardeau-Montaut, 2016), and then integrated into Schlumberger's *Petrel* modeling software (Fig. 4.6b), as the basis for interpretation. In order to establish a digital outcrop model (Fig. 4.6c), the coal seams of the cyclic successions (Figs. 4.4, 4.6a) were used as markers to establish the major horizons, based on their characteristic low reflection intensity (Fig. 4.5c). The interpretation was extended with measured bedding orientation data and historical mining data (Haarmann, 1909), besides the interpreted horizon traces in the digital outcrop model, for a more reliable surface reconstruction of the westerly-plunging anticline. Two major W-E striking normal faults, with an offset of > 10 m (83 ft), were mapped in the digital outcrop model. Data was extended by manual orientation measurements in the outcrop because, in most cases, only fault traces can be mapped in the digital outcrop model. Horizons were modeled, accounting for both the fault model and reconstructed surfaces, with a focus on the main sandstone bodies of the cyclic successions. Zones were created for major facies units. Literature data were used to reconstruct the thickness of coal seam horizons prior to mining activities (Haarmann, 1909).

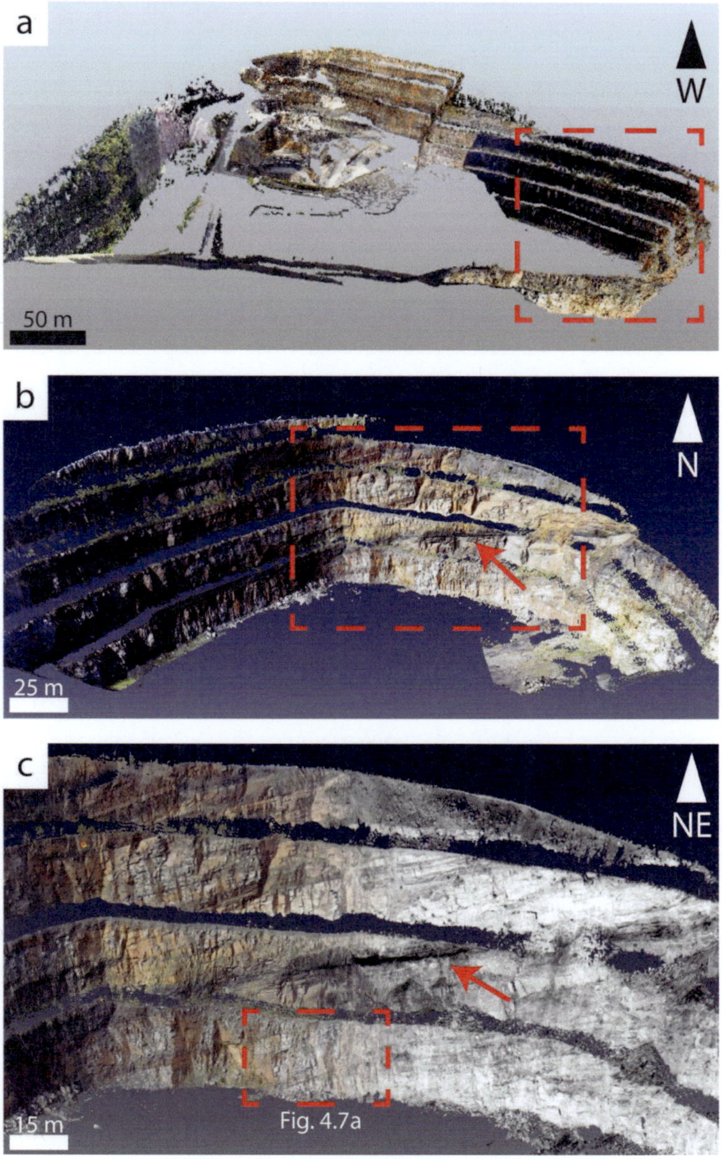

Figure 4.5: Georeferenced 3D t-LiDAR point clouds of the quarry. a) View from the west into the quarry (true color-coding). b) Magnification of the NE part of the quarry in true color-coding, indicated by the dashed box in (a). c) Magnification of the dashed box in (b). The left side of the image is true color-coded while the right side is gray color-coded, which correlates with the intensity of the reflected laser beam. The red arrows point to the "Zweibänke" coal seam, which is represented in relatively dark colors, compared to the bright surrounding sandstone units. Both color codings represent valuable data for the differentiation of lithologies.

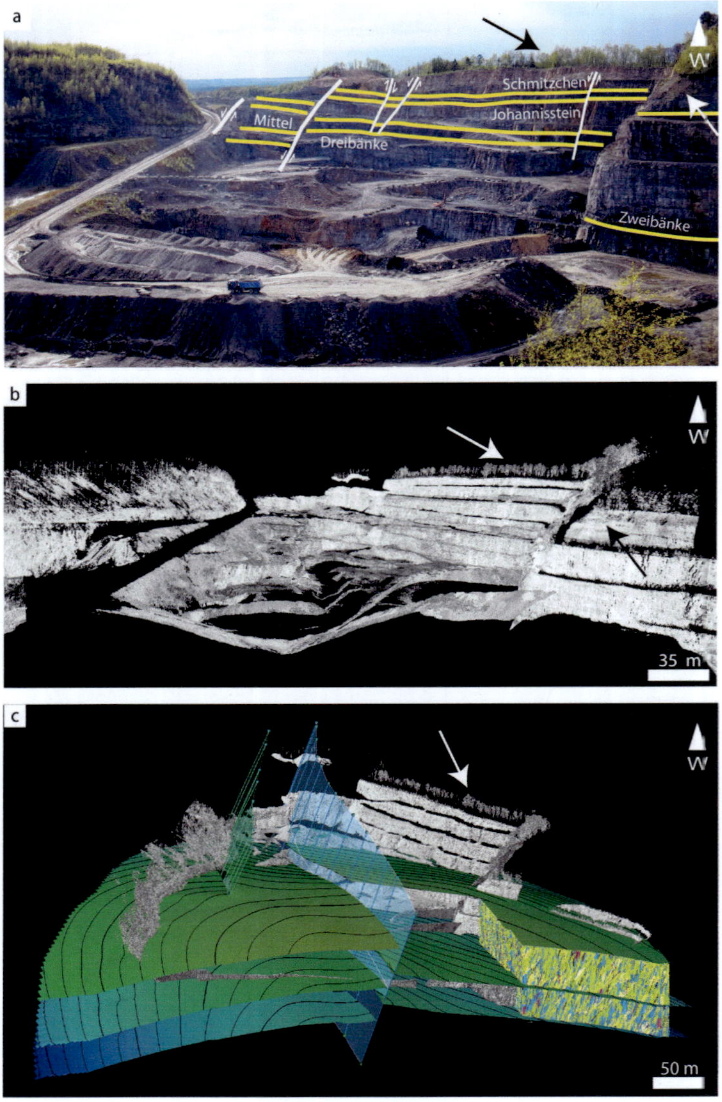

Figure 4.6: Workflow from outcrop to 3D outcrop analog model. All images are the same view of the quarry, facing west. a) Photo image of the quarry with indicated coal seams and W-E striking normal faults. Note the truck for scale (after Wüstefeld et al., 2017a). b) Merged point cloud of the entire quarry from laser scans imported into Schlumberger's reservoir modeling software Petrel as the basis for interpretation and 3D geometry reconstruction. c) Established 3D geometries comprising two representative W-E striking normal faults (both left faults in (a)) and main horizons (coal seams in (a)), as well as the modeled DFN in the test area for two different stratigraphic positions. The arrows pointing to the right in all pictures highlight the same trees atop the quarry for scale and orientation. The remaining arrows, pointing to the left, mark the same large-scaled edge in the quarry.

4.4.3 Automated discontinuity surface detection

The t-LiDAR point cloud data represents the entire 3D morphology of outcrop surfaces, including prominent discontinuity surfaces, such as fractures and bedding (Fig. 4.7a). For the presented workflow, it was assumed that fracture surfaces are planar (Fig. 4.6b; e.g. Pollard and Aydin, 1988). The developed algorithm relies on a bottom-up/agglomerative, hierarchical clustering approach (Johnson, 1967) and was specially adapted for fracture detection requirements. Such methods start with a large number of individual points, which are successively merged into clusters, according to a well-defined similarity criterion, representing the sought planar regions (Fig. 4.7c). To initialize the clustering approach, a normal is associated with each point (Fig. 4.7c), which is done by performing principal component analysis (PCA), based on covariance matrices (Hotelling, 1933) in a small neighborhood of each point. The PCA is similar to the orientation tensor of Woodcock (1977), when neglecting the conversion of the point set barycenter to the origin of the coordinate system (cf. Seers and Hodgetts, 2016b). The total neighboring points were set from 60 to 100 for different scanlines. The resulting eigenvalues of the PCA mathematically represent each point neighborhood as an ellipsoid. If the smallest axis of this ellipsoid (i.e. the smallest eigenvector) is reasonably small, compared to its other axes (curvature criterion: the user-defined input is set to a maximum of 0.04 for the ratio of the smallest axis, divided by the square root of the product of the remaining axis), a small, flat surface area is encountered. For such a point a cluster is initialized and the normal of the cluster defined to be the smallest axis of the ellipsoid. This procedure fits a plane to the point neighborhood, where the point normal equals the plane normal. During cluster initialization, surface areas are omitted and not used as initial clusters, which do not match the curvature criterion, i.e. where the smallest axis is of a similar magnitude to the other axes of the ellipsoid. This prevents the formation of micro-clusters, for example at the edges of intersecting fracture surfaces. The agglomerative hierarchical clustering algorithm then successively merges spatially neighboring clusters with the highest similarity of their normal to a single, larger cluster. After each merge, a new normal is computed for the new point subset. The algorithm stops if the similarity of all neighboring clusters is above a user-defined angle threshold, which was set to a maximum of 15 for this study. Now, previously neglected clusters, as for example corner points, are added to those clusters, where point to plane distances are minimal and below a user-defined threshold of 1 cm. In this study, we defined a minimum size threshold of 25 points for the detected clusters (fracture surfaces). This corresponds to a size of 0.02 m^2 in 50 m and of 0.09 m^2 in 100 m distance from the laser scanner (cf. distance-dependent resolution of point cloud in the outcrop data acquisition section, above). An elliptical plane is then fitted in each identified cluster by the algorithm, which represents both orientation and spatial position (Fig. 4.7c). Automated discontinuity surface detection was implemented in the open source geometry processing and rendering framework *OpenFlipper* (Möbius and Kobbelt, 2012).

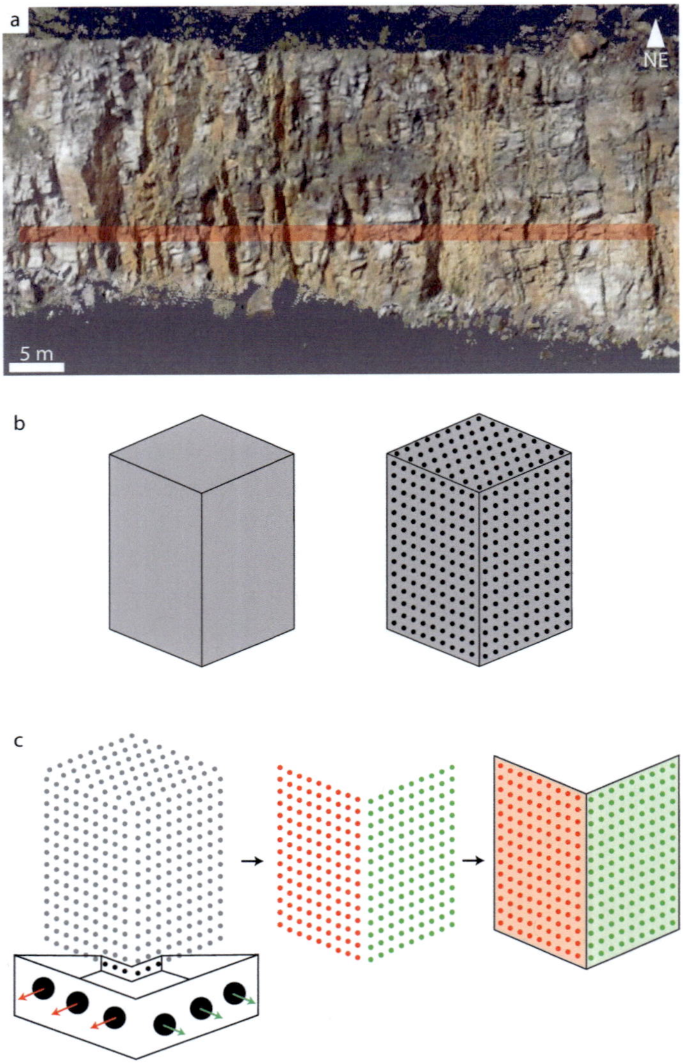

Figure 4.7: Automated discontinuity surface detection. a) Magnification of box in Fig. 4.5c, highlighting the 3D morphology of the quarry wall, with gently-inclined bedding, and subvertical orthogonal fracture surfaces truncating the complete sandstone cycle. The red-colored area symbolizes a virtual scanline, along which surfaces are extracted to serve as the basis for the fracture analysis. b) Cuboid representing simplified fracture surfaces. This geometric object is technically represented by individual points. c) Schematic approach of fracture discontinuity surface detection by the developed algorithm. Simplified, for each point the orientation is calculated (colored arrows) with respect to neighboring points. This is the primary criterion for the subsequent clustering to detect points that belong to the same planar surface. Eventually, an ellipse (represented by a plane) is fitted into the cluster, which represents the spatial information of the detected surface.

4.4.4 Virtual scanlines as basis for fracture analysis

The presented workflow for fracture analysis is closely aligned to standard hydrocarbon exploration and production industry subsurface workflows based on 1D well log data (e.g. borehole image logs and cores). Therefore, the fracture analysis relies on the establishment of virtual scanlines (*sensu* Voyat et al., 2006) in the digital 3D outcrop model, which can also be termed virtual (lateral) boreholes, as they technically represent the same. This concept allows for verification and comparison with conventional manual data based on scanline measurements of the outcrop. Automated discontinuity surface analysis, using the developed algorithm, relies on extracted point clouds along defined scanlines. In this study, we used a width of 0.5 – 1 m, but this can be easily adjusted to fit the individual case study, i.e. fracture characteristics and is therefore not restricted to the given values. The user can create these point clouds, representing virtual scanlines, at any suitable position of the 3D outcrop model (Fig. 4.8) by defining the position and length on the virtual quarry walls. The output of the automated fracture discontinuity surface detection are ellipsoids that bear both the spatial position as well as orientation of the fracture surfaces along the investigated virtual scanline. The architecture of the algorithm, in combination with the interpretation along virtual scanlines, permits very short computational times, i.e. a couple of seconds. Data integration into Schlumberger's reservoir modeling software *Petrel* is realized by the establishment of (lateral) well trajectories and the incorporation of the detected fracture planes along this trajectory (Fig. 4.8). The extracted fracture planes can be quantified and visualized in well logs (cf. section on fracture analysis, below). A back-tilting of obtained fracture data was not applied as the fracture data is aimed to be used for modeling purposes, i.e. the data shall be represented as detected. The bedding of the quarry dips only gently with 15° to the NNW and with 20° to the S, on the northern and southern limbs, respectively. This would cause a rotation of the strike of the fracture poles by less than 2° and a variation of the dip of less than 15°.

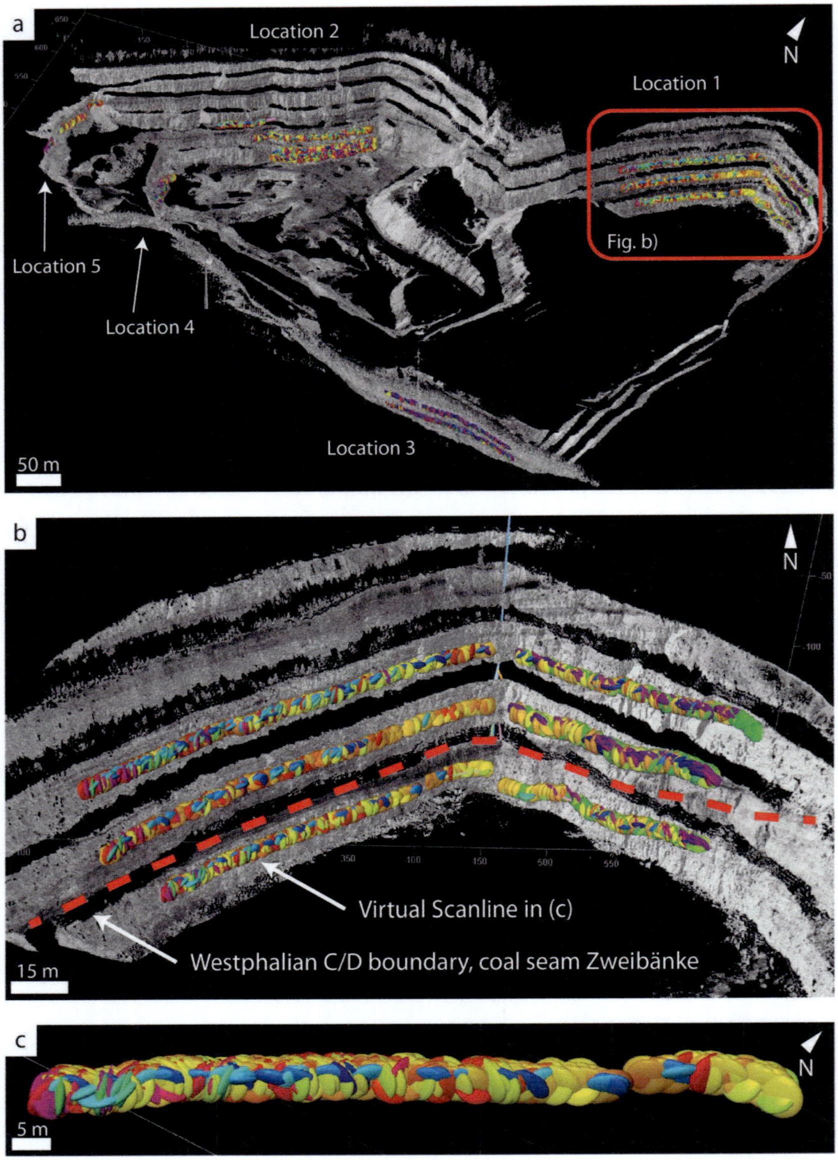

Figure 4.8: a) 3D point cloud of the entire Piesberg Quarry, implemented in Schlumberger's Petrel reservoir modeling software. The investigation areas are indicated (cf. Fig. 4.3), for which virtual scanlines were established (cf. Fig. 4.7). These scanlines serve as the basis for the fracture analysis workflow of this study, and can be established in any suitable position of the digital outcrop model. The detected fractures are implemented along (lateral) well trajectories at corresponding positions of the previously extracted virtual scanlines. b) Magnification of box in (a). The detected fractures are represented by flat spheres, which are color-coded according to dip azimuth. c) Magnification of a single virtual scanline as indicated in (b).

4.4.5 Fracture analysis

The virtual well trajectories, including the corresponding well logs, are exported to the fracture modeling software *FracaFlow* (Fig. 4.9; a product of Beicip-Franlab). In a first step, the well logs were analyzed with respect to principal fracture orientations. In order to describe the spatial distribution of the detected fractures, we followed the P_{xy} scheme (Dershowitz, 1985; Dershowitz and Herda, 1992), where P_{10} data represents the fracture count per unit length (Fig. 4.9c). For the virtual scanline, comprising all detected fractures as well as for each identified fracture set, based on orientation statistics, the mean P_{10} fracture density and P_{10} density logs were derived from the established well logs using the fracture density log computation module of the *FracaFlow* software (Fig. 4.9b). The P_{10} data is orientation-dependent and needs to be corrected with respect to the virtual scanline orientation, i.e. the measured direction, based on the principle of Terzaghi (1965). Fracture sets with angles of less than 30° between the strike of the fracture set and virtual scanline orientation should be excluded from fracture density estimation as the Terzaghi correction can result in significant overestimations (e.g. Bisdom et al., 2014). If possible, orthogonal scanline orientations were chosen to cover the full set of fracture orientations.

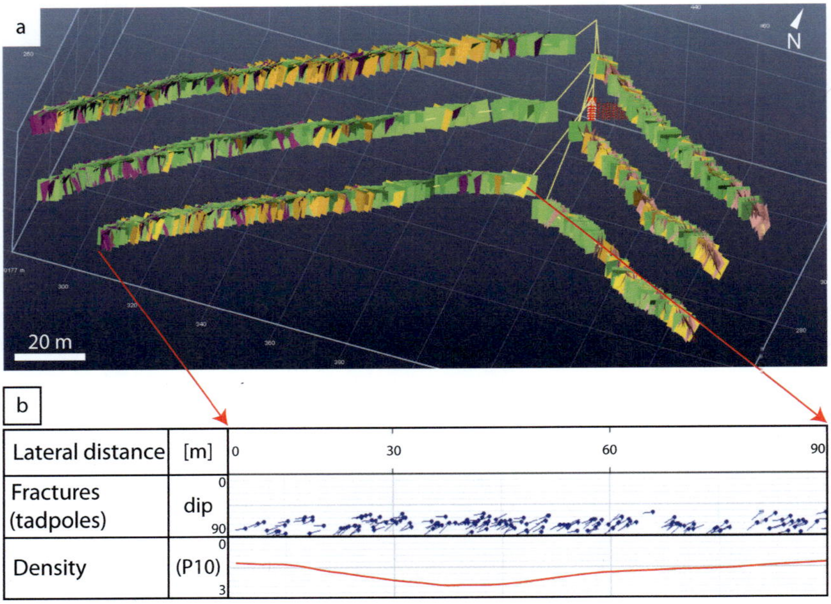

c

Measurement dimension → / Object / Fracture dimension ↓	Length 1D	Area 2D	Volume 3D
Number of fractures (0)	**P 10** Number of fractures per length unit [m^{-1}]	**P 20** Number of fractures per surface unit [m^{-2}]	**P 30** Number of fractures per volume unit [m^{-3}]
Fracture length (1)		**P 21** Cumulative fracture length per surface unit [m^{-1}]	**P 31** Cumulative fracture length per volume unit [m^{-2}]
Fracture area (2)		**P 22** Fracture area per volume unit [%]	**P 32** Cumulative fracture area per volume unit [m^{-1}]

Figure 4.9 (Description on next page)

Figure 4.9 (Previous page): a) The virtual lateral well trajectories, including the derived fractures, are exported to Beicip's fracture modeling software FracaFlow for fracture orientation and density analysis. b) Computed density log for the ENE-WSW fracture set, based on the 1D virtual scanlines, which contains the tadpoles and P_{10} fracture density. c) The P_{xy} scheme (Dershowitz, 1985; Dershowitz and Herda, 1992), which describes the spatial distribution of fractures. The virtual scanlines represent P_{10} data and can be converted to P_{32} data based on stereological assumptions (e.g. Dershowitz, 1985; Dershowitz and Herda, 1992; Wang, 2005), which is needed for stochastic DFN modeling in the scope of this study.

4.4.6 Discrete fracture network model

The creation of a stochastically modeled DFN aims to demonstrate the transferability of digital mapping data into suitable data for modern reservoir modeling software. The derived 1D fracture density data (P_{10}, cf. Fig. 4.9c) need to be converted, based on stereological assumptions, to obtain required P_{32} values for 3D static DFN modeling (e.g. Dershowitz, 1985; Dershowitz and Herda, 1992; Wang, 2005). In this study, the conversion was realized with Beicip's fracture modeling software *FracaFlow* (cf. Barthélémy et al., 2009); however, such transformations might introduce uncertainties and consequently need to be handled with care. Bisdom et al. (2014), for example, underscored the necessity to carefully define the fracture sets since the conversion from P_{10}/P_{21} to P_{32} data might differ with respective different orientation characteristics. The DFN was generated with the fracture-modeling module of Schlumberger's *Petrel* reservoir simulation software. The previously created 3D model served as a basis in which the DFN was modelled in a test area (Location 1, Figs. 4.3, 4.8a) with an approximate lateral dimension of 90 x 180 m and a 50 m height (Fig. 4.6c). Each fracture set was modeled individually for two zones, representing different stratigraphic positions, which are separated by a coal seam (Fig. 4.8a). The test area was chosen because of the largest amount of individually investigated fractures (cf. Fig. 4.3). Since reasonable lateral voxel sizes in standard industry approaches are about 20 – 25 m wide, the density of the derived data is too high for standard modeling approaches. Therefore, the input parameters for the DFN modeling include the derived mean dip azimuth and dip direction (Fisher distributed) for each fracture set as well as the averaged P_{32} fracture intensities. The workflow does not allow the direct derivation of fracture size and size distribution. For this reason, a log-normal length distribution, with a 25 m mean fracture length and constant aspect ratio of 2 (e.g. Odling et al., 1999; Hitchmough et al., 2007) as well as polygons with 20 sides, were assumed for this study. The maximum fracture size is thereby limited by the extent of the modeled domains. After the automated fracture detection and data integration into the reservoir modeling software, the conventional industrial workflow can be applied to the DFN modeling. The outlined approach is not restricted to the software solutions that are utilized in this particular study and can be substituted by other equivalent programs. The entire workflow from outcrop data acquisition over the digital fracture analysis to the modelling processes is presented in the flowchart in Fig. 4.10.

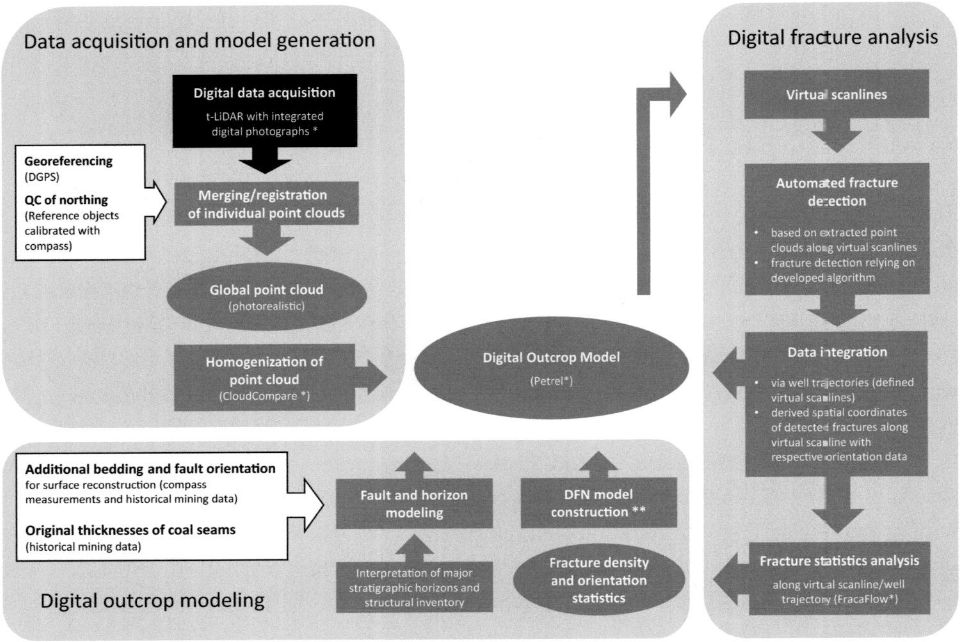

Figure 4.10: Flowchart representing the outlined workflow from outcrop analog to digital outcrop model, including data acquisition and model generation, digital outcrop modeling and digital fracture analysis. NB *The technology/software used in this case study can be substituted with equivalent technology/software solutions. **For the construction of a DFN, further parameters (fracture size and aperture) need to be derived or estimated.

4.5 Results

4.5.1 Fracture analysis of synthetic scanline

The presented method was applied to a synthetic virtual scanline to test the validity of the approach. The artificial section, a 3D point cloud with 5600 points in total and 10 cm distance between neighboring points, consists of planar surfaces with varying orientations and has a total length of 43 m and a width of 1 m. Furthermore, a random noising of up to 6 cm deviation was incorporated for each point. The digital analysis detected 57 of the 58 fractures (Fig. 4.11) and delivered similar orientation (Fig. 4.11a) and fracture spacing distributions (Fig. 4.11b).

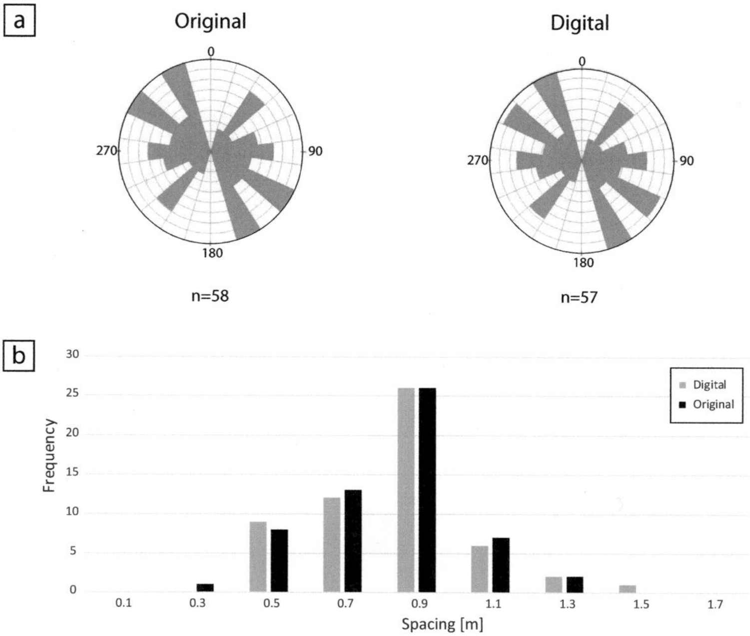

Figure 4.11: Results of the digital analysis of the synthetic virtual scanline (digital) in comparison with the data, which was used to create the synthetic virtual scanline (original). a) Strike diagrams representing the orientation analysis. b) Spacing frequency distributions representing the fracture density analysis.

4.5.2 Fracture analysis of case study

In the virtual 3D model, a total of 17 virtual scanlines, at five different outcrop locations, were investigated, covering different stratigraphic and lithological positions (Fig. 4.8a). The virtual scanlines were chosen to reveal relationships between fracture pattern and stratigraphy and distance to faults, as well as to compare quarry walls with different mining stages. In the scope of this study, only discontinuity surfaces with a dip angle of more than 50° were considered to exclude sedimentary bedding and cross-bedding. In total, 3017 investigated discontinuity surfaces, along a total length of investigated scanlines of approximately 1620 m, were detected.

4.5.2.1 Comparison of manual and digital measurements

In the northeast of the quarry (Location 1, Figs. 4.3, 4.8a) six virtual scanlines were measured, established within two sub-perpendicular oriented quarry walls (Figs. 4.3, 4.8b), comprising a total of 1336 detected fracture surfaces. Two main fracture sets (WSW-ENE and NNW-SSE) can be deduced from the virtual data (Fig. 4.12a, b), and their orientation is comparable to manual compass measurements (Fig. 4.12c). Compass data were taken below the “Zweibänke” coal seam (Fig. 4.4) on both orthogonally-oriented quarry walls at Location 1 (Fig. 4.3). For the traditional fracture-spacing assessment, manual scanlines were measured at similar positions at the lowermost sections of both perpendicularly-oriented quarry walls at Location 1 (Figs. 4.3, 9a); however, due to accessibility, the manual measurements were conducted over a shorter distance of 30 m, compared to the length of the virtual scanlines of 70 – 120 m. The spacing frequency distributions of the manual scanline measurements are comparable for both quarry walls with the automated fracture analysis (Fig. 4.13), revealing an average fracture spacing of 0.38 m/0.55 m (manual scanlines) *vs.* 0.45 m/0.54 m (digital scanlines) of the NE-SW and NW-SE oriented quarry walls, respectively.

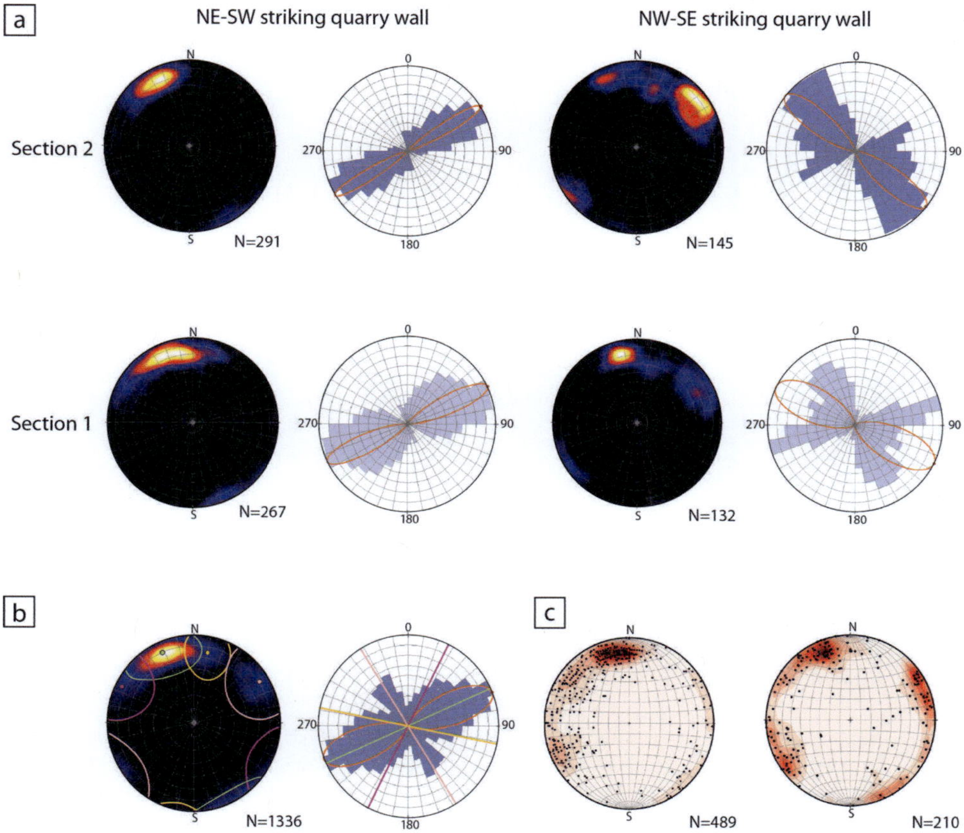

Figure 4.12: Fracture orientation analysis of Location 1 (Figs. 4.3, 4.8a) in the NE of the quarry. a) Stereonet and strike diagrams of two sub-perpendicularly-oriented quarry walls and two different stratigraphic positions. Section 1 was sampled below the "Zweibänke" coal seam (Fig. 4.8b), while Section 2 was sampled in the hanging wall of the same. b) Stereonet and strike diagram of the entire Location 1, including all individually-sampled virtual scanlines. Four fracture sets were identified, comprising two sub-orthogonal fracture orientations, as indicated in the strike diagram. c) Stereonet based on conventionally-established manual measurements from Location 1, lower section (compare with automated fracture data in (a), Section 1). The left diagram represents the NE-SW oriented quarry wall, while the left diagram is the NW-SE oriented quarry wall.

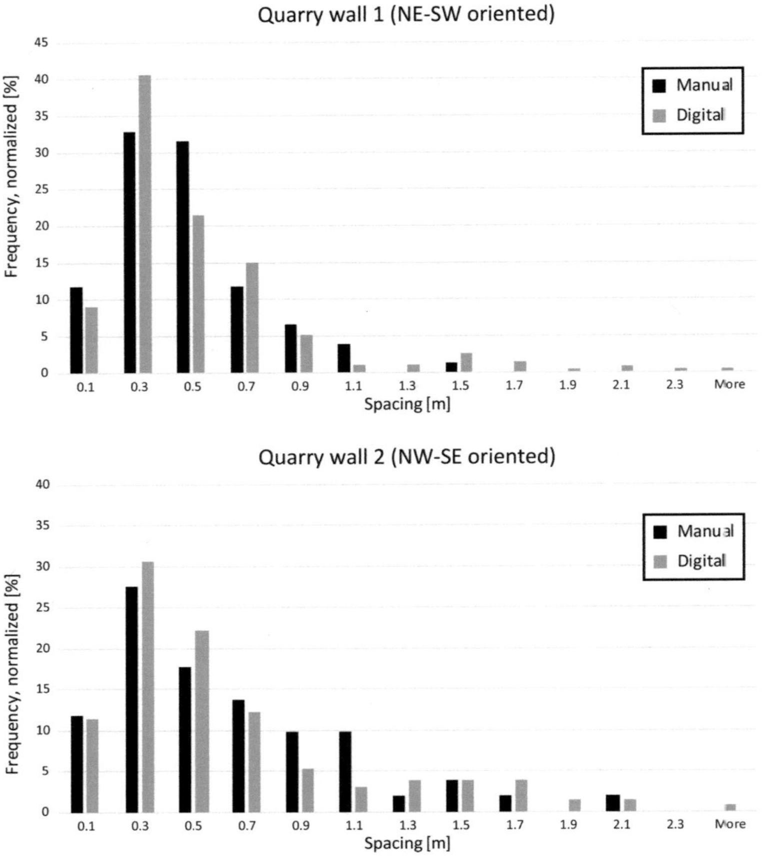

Figure 4.13: Comparison between fracture spacing frequency distributions, based on the digital analysis along virtual scanlines, and manual measurements along scanlines for the lowermost sections of both perpendicularly-oriented quarry walls of Location 1 (Fig. 4.3). The manual data was taken along scanlines with similar positions, but over shorter distances, compared to the virtual scanlines.

4.5.2.2 Fracture orientation analysis

In total, four fracture sets, comprising two sub-orthogonal fracture orientations (Fig. 4.12b), were identified, based on the individual and compound analysis of all six virtual scanlines of Location 1 (Figs. 4.3, 4.8a). On the NE-SW oriented quarry wall, the WSW-ENE striking fracture set is dominant, while in the NW-SE striking quarry wall the WSW-ENE as well as the NNW-SSE to NW-SE striking fracture sets prevail (Fig. 4.12a). Below the "Zweibänke" coal seam (Section 1, Fig. 4.12a), the NNW-SSE striking fracture set is dominant.

Sections 1-4 of the second sampling site on the northern quarry wall (Location 2, Figs. 4.3, 4.8a, 14) represent a coarse- to medium-grained fluvial facies, and have an overall comparable main fracture orientation. Section 5 (Fig. 4.14) is a fine- to medium-grained sandstone of a different facies (lacustrine delta). Here, the orientation pattern varies and has a slight clockwise rotation in orientation (Fig. 4.14) as opposed to Sections 1 to 4. The lithological descriptions and variation are based on outcrop studies and direct work on the quarry faces (cf. Wüstefeld et al., 2017a).

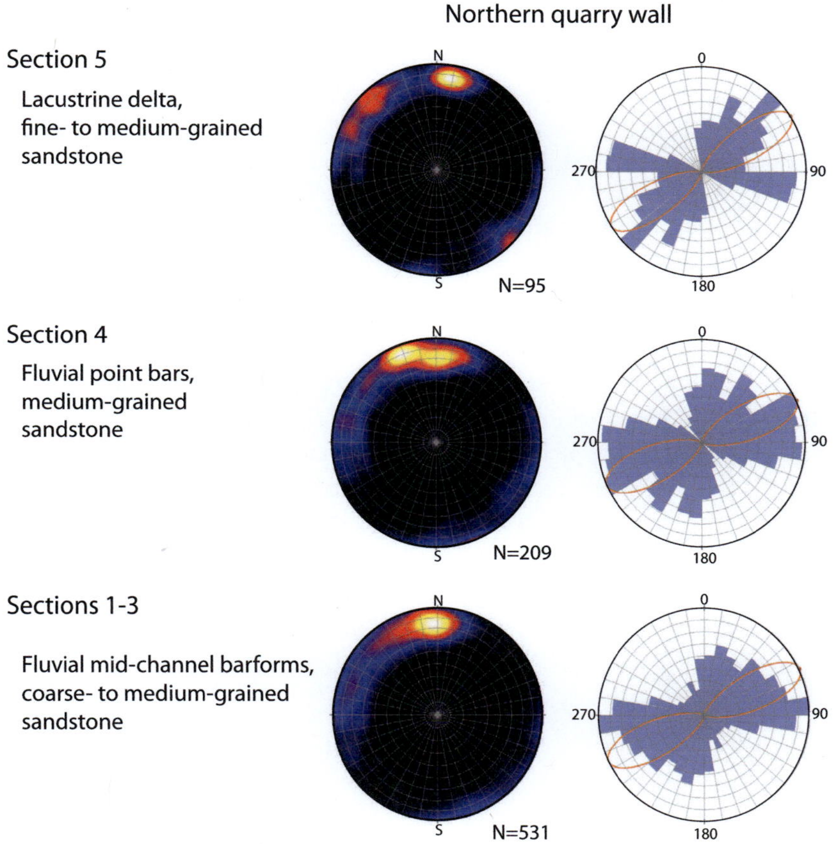

Figure 4.14: Stereonet and strike diagrams of fractures measured along five lateral profiles of 100 – 125 m length at Location 2 (Figs. 4.3, 4.8a). Each stereonet and strike diagram represents a different stratigraphic position. Sections 1-3 (at the base) are coarse- to medium-grained sandstones and Section 4 is a medium-grained sandstone, both of which can be attributed to a fluvial facies. Fractures of Section 5 are from a fine- to medium-grained sandstone with a lacustrine delta origin. The lithological description is based on outcrop studies (cf. Wüstefeld et al., 2017a).

In order to compare and derive lateral fracture strike changes across the quarry, all detected fractures of the individual locations (Figs. 4.3, 4.8a) were plotted on a single stereonet for each site (Fig. 4.3). The WSW-ENE striking fracture set is dominant in the northeastern part of the quarry, which is closest to the large NNW-SSE striking normal fault in the east (Location 1, Figs. 4.3, 4.8a), with an offset of up to 600 m. Further away from this major normal fault, the primary fracture orientation slightly rotates clockwise to a W-E strike, which can be deduced by comparing Locations 1 to 3 (Fig. 4.3). In contrast, a WNW and NW strike of fractures measured along a NNW-SSE oriented quarry wall in the western quarry are consistent with measurements along a NW-SE oriented quarry wall in the east (Fig. 4.3, Location 4, Fig. 4.12a); however, scanlines of Location 4 and 5 were both taken in the vicinity (< 5 m distance from fault zone) of a W-E striking normal fault (f 1 in Fig. 4.3), with a displacement of approximately 15 m. Three smaller-scale faults, with a throw of < 5 m additionally cross the scanline of Location 5. Both locations have, in general, a distinct fracture orientation pattern (Fig. 4.3).

4.5.2.3 Fracture density analysis

The mean P_{10} densities (corrected and uncorrected, Tab. 4.1) of all identified fracture sets at Location 1 (Figs. 4.3, 4.8a) highlight the same dominance of fracture sets for both perpendicularly-oriented quarry walls as already shown previously in the fracture orientation analysis (Fig. 4.12). Furthermore, the densities reflect the same change of dominance of the NNW-SSE striking fracture set of quarry wall 2 (NW-SE oriented), when comparing the stratigraphic units separated by the “Zweibänke” coal seam (Fig. 4.12a, between Sections 1 and 2). The uppermost section at Location 1 and the lowermost Sections at Location 2 (Sections 1-3) belong to the same stratigraphic position. The corresponding fracture sets have a comparable density (Tab. 4.1). While Section 1-4 of Location 2 (Tab. 4.1) have an overall similar fracture density and represent a coarse- to medium-grained fluvial facies (Wüstefeld et al., 2017a), Section 5 is a fine- to medium-grained lacustrine facies and has an increased density of the NNE-SSW striking fracture set (Tab. 4.1).

Table 4.1 (Description on next page)

Location 2						Location 1									
							Quarry wall 1 (NE-SW oriented)					Quarry wall 2 (NW-SE oriented)			
Fracture Set	Scanline	P 10 [1/m] (uncorrected)	SD	P 10 [1/m] (corrected)	SD	Fracture Set	Scanline	P 10 [1/m] (uncorrected)	SD	P 10 [1/m] (corrected)	SD	Scanline	P 10 [1/m] (uncorrected)	SD	P 10 [1/m] (corrected)	SD
WSW-ENE	*Section 5*	1.0	0.2	1.7	0.5	**WSW-ENE**	Section 3	1.3	0.5	2.5	1.0	Section 3	0.7	0.2	0.8	0.2
	Section 4	1.0	0.1	2.0	0.3		Section 2	1.7	0.6	3.4	1.2	Section 2	0.4	0.1	0.5	0.2
	Section 3	1.1	0.4	2.2	0.8		Section 1	1.6	0.5	3.1	1.1	Section 1	0.8	0.3	1.1	0.5
	Section 2	1.0	0.3	1.9	0.7											
	Section 1	1.1	0.1	2.1	0.4	**NNE-SSW**	Section 3	0.6	0.5	1.2	0.9	Section 3	-	-	-	-
							Section 2	0.3	0.2	0.5	0.5	Section 2	-	-	-	-
NNE-SSW	*Section 5*	0.9	0.1	1.7	0.3		Section 1	0.3	0.3	0.6	0.5	Section 1	-	-	-	-
	Section 4	0.5	0.2	0.9	0.5											
	Section 3	0.5	0.2	0.9	0.3	**NNW-SSE**	Section 3	0.1	0.1	0.1	0.1	Section 3	1.1	0.3	2.1	0.5
	Section 2	0.6	0.1	1.0	0.3		Section 2	0.02	0.03	0.02	0.04	Section 2	1.3	0.4	2.5	0.7
	Section 1	0.6	0.2	1.2	0.4		Section 1	0.03	0.1	0.03	0.1	Section 1	0.8	0.3	1.6	0.6
NNW-SSE	*Section 5*	0.1	0.03	0.1	0.04	**WNW-ESE**	Section 3	0.4	0.2	0.6	0.4	Section 3	0.2	0.1	0.4	0.2
	Section 4	0.2	0.1	0.2	0.2		Section 2	0.1	0.1	0.1	0.1	Section 2	0.4	0.2	0.8	0.4
	Section 3	0.3	0.1	0.3	0.2		Section 1	0.4	0.2	0.6	0.4	Section 1	0.3	0.2	0.6	0.4
	Section 2	0.2	0.1	0.3	0.2											
	Section 1	0.2	0.04	0.2	0.1											

Table 4.1 (Previous page): Fracture density analysis of Location 1 and 3 (Figs. 4.3, 4.8a) for each virtual scanline. For Location 1, two sub-perpendicular quarry walls were investigated. The quarry wall orientation of Location 2 is comparable to quarry wall 1 of Location 1. The stratigraphic position represented by Section 3 of Location 1 is similar to those of Sections 1-3 of Location 2. The P10 fracture density (uncorrected and corrected, cf. Fig. 8c), including the standard deviation (SD), is derived for the main fracture sets. Sections 1-3 (at the base) of Location 2 are coarse- to medium-grained sandstones, and Section 4 is a medium-grained sandstone, both of which can be attributed to a fluvial facies. Fractures of Section 5 are from a fine- to medium-grained sandstone of a lacustrine delta. The lithological description is based on outcrop studies (cf. Wüstefeld et al., 2017a).

Comparison of the individual fracture sets across the quarry is not favorable, as it was shown that: a) the dominance of the primary fracture orientation can vary even over short distances (as explained in the following section, cf. Fig. 4.12); b) dominant fracture sets at the same location vary due to lithological changes (Fig. 4.14); and c) a clockwise rotation of the main fracture sets in westerly direction in the quarry can be seen (Locations 1-3, Fig. 4.3). The evaluation of the locations across the quarry, with overall comparable main fracture orientations, demonstrates that overall average fracture spacing is similar throughout the quarry, ranging from 0.42 to 0.55 m (Fig. 4.15). Consequently, the resulting mean total fracture count per unit length (uncorrected and corrected) is homogeneous throughout the quarry (Tab. 4.2).

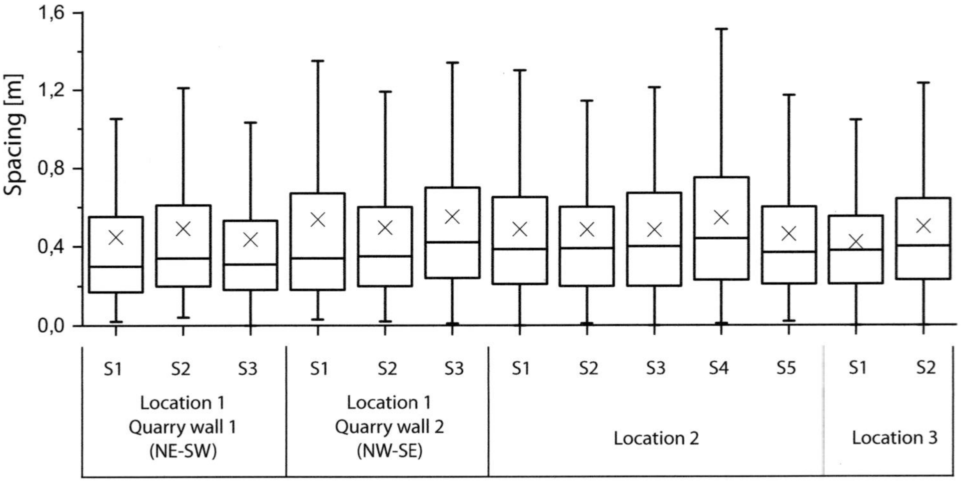

Figure 4.15: Box-Whisker-plots of fracture spacing (all detected fractures), comparing sampling sites across the quarry with overall comparable main fracture orientations (Locations 1-3).

Table 4.2: Comparison of P_{10} fracture density (uncorrected and corrected, derived for all detected fractures) for sample sites across the quarry with overall comparable main fracture orientations (cf. Fig. 4.15).

			P 10 [1/m] (uncorrected)	SD	P 10 [1/m] (corrected)	SD
Location 3		Section 2	2.2	0.4	4.1	0.9
		Section 1	2.4	0.4	4.5	0.9
Location 2		Section 5	2.2	0.2	4.0	0.5
		Section 4	1.9	0.3	3.4	0.5
		Section 3	2.1	0.5	3.9	0.8
		Section 2	2.1	0.6	3.9	0.8
		Section 1	2.2	0.7	4.0	0.7
Location 1	Quarry wall 1 (NE-SW)	Section 3	2.4	0.9	4.4	1.7
		Section 2	2.1	0.8	4.1	1.5
		Section 1	2.3	0.9	4.4	1.7
	Quarry wall 2 (NW-SE)	Section 3	1.9	0.5	3.3	0.6
		Section 2	2.1	0.8	3.8	1.1
		Section 1	1.9	0.8	3.3	1.2
		Mean	2.1	0.6	3.9	1.0

Laterally corrected density logs for Sections 1-4 at Location 2 (Figs. 4.3, 4.8a) are presented for all fractures as well as exemplarily for the WSW-ENE striking fracture set (Fig. 4.16). The logs show a high lateral heterogeneity of the fracture density over short distances of 10 m from a P_{10} of 1 m^{-1} to a P_{10} of 7 m^{-1}. The overall trends of Sections 1-3, with a mean corrected fracture density of 3.9 m^{-1} (all fractures, Fig. 4.16), have a comparable shape with zones of increased fracture densities some 10 to 20 m wide. In contrast, the shape of the logs of Section 4, with a mean fracture density of 3.4 m^{-1} (all fractures, Fig. 4.16), have different highs and lows.

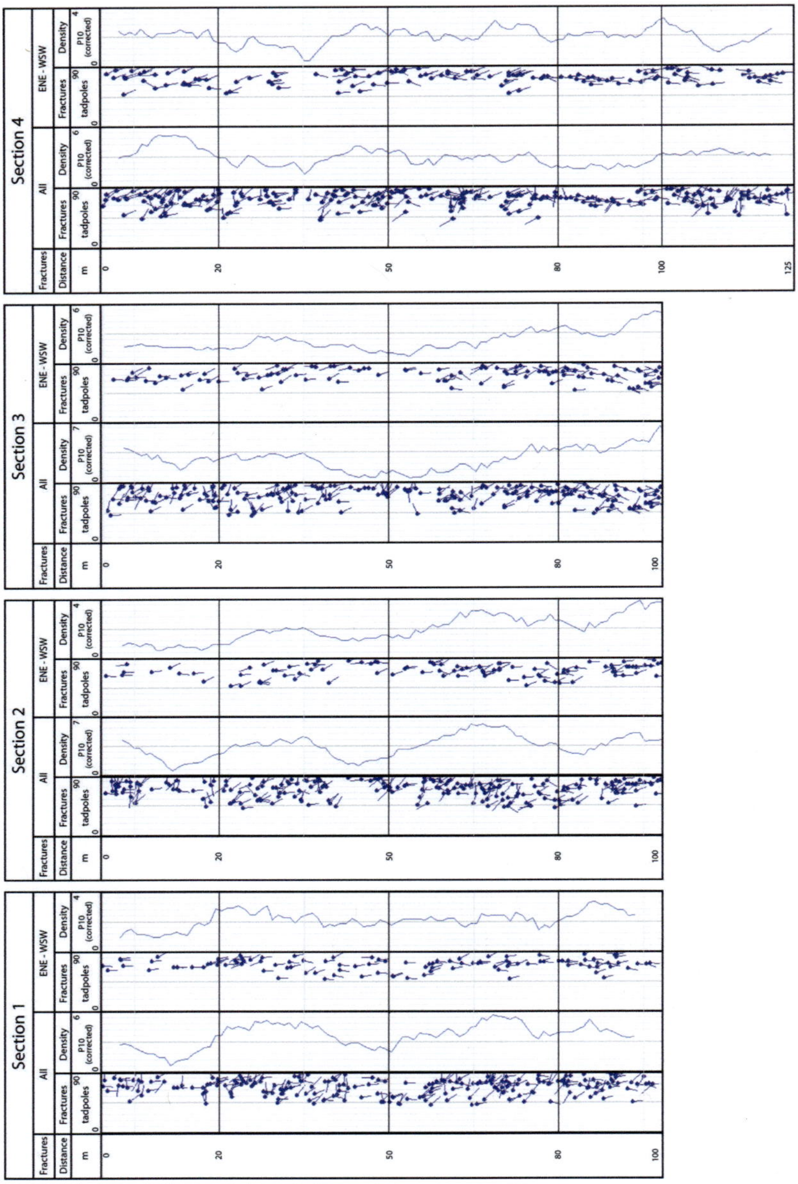

Figure 4.16: Lateral fracture density logs of different sections of Location 2 (Fig. 4.3) for all identified fractures as well as exemplarily for the WSW-ENE striking fracture set. For comparison, all logs are aligned to equivalent lateral positions. Distance 0 is located on the eastern side. All sections can be attributed to a similar stratigraphic position. While Sections 1-3 are vertically closely aligned within a 20 – 30 m thick unit, representing coarse- to medium-grained fluvial sandstones, Section 4 represents a different lithotype of medium-grained fluvial sandstones (cf. Wüstefeld et al., 2017a). All profiles are crossed by a minor fault with < 2 m displacement at approximately distance 10 in the logs.

4.6 Discussion

4.6.1 Fracture analysis

4.6.1.1 Comparison of manual and digital measurements

The principal fracture orientations (WSW-ENE and NNW-SSE) at Location 1 were identified by both digital and manual analysis. Due to the more significant amount of investigated fracture surfaces and the incorporation of areas inaccessible for manual measurements, further fracture sets with WNW-ESE and NNE-SSW strikes for the entire Location 1 were derived by digital analysis (Fig. 4.12). The small differences in the fracture spacing distribution between digital and manual derivation, besides the overall similar results, are likely the result of differences in the length of the manual and virtual scanlines. Thus, the virtual scanlines provide a statistically more significant dataset.

4.6.1.2 Fracture density and orientation

The deduced zones of highly clustered fractures for Sections 1-3 (Fig. 4.16), some 10 to 20 m wide (Fig. 4.16), can be correlated across 20 to 30 m thick units and might be interpreted as fracture corridors; however, the data did not reveal a relationship between the westerly-striking fault at Location 2 (f 2, Fig. 4.3) on the northern wall, with a maximum 2 m displacement, and the increased fracturing or rotation of fractures. Section 4 (Fig. 4.16) belongs to the same fluvial cyclic deposit but represents a different lithotype of medium-grained sandstones. Here, zones of increased fracturing have similar dimensions; however, the different positions of the highs and lows (Fig. 4.16) might reflect a lithological control. Furthermore, the individual fracture set density and orientation vary vertically for different facies/lithotypes (e.g. Tab. 4.1, Location 2, Sections 4 and 5; Fig. 4.14), which may be due to small differences in the elastic properties of the different lithotypes (e.g. Gross et al., 1995; Bourne, 2003).

The derived principal fracture orientations are in accordance with previous studies, which describe a main NNW-SSE fracture orientation for the entire quarry as well as a WNW-ESE striking set, the latter absent in the eastern part of the quarry, where WSW-ENE fractures prevail (Drozdzewski, 1998; Bissen, 2011). Our vast dataset also highlights the dominance of the WSW-ENE fracture set in the eastern quarry, but also a minor WNW-ESE set.

Contrarily to the fracture pattern around the westerly striking, small-scale fault (displacement < 2 m), fractures vary laterally in strike direction across the study area with a clockwise rotation of the main fracture sets from WSW-ENE towards W-E in a westerly direction (Locations 1-3, Fig. 4.3). Considering that the large NNW-SSE striking normal fault in the east truncates the study area (e.g. Figs. 4.3, 4.4), the local stress field around such faults may have caused the rotation of the fracture patterns. Stress shadows around faults are frequently reported to cause stress rotation and associated rotation of fractures to become perpendicular to the fault strike (e.g. Rawnsley et al., 1992; Finkbeiner et al., 1997; Bourne and Willemse, 2001; Yale, 2003); however, an increased fracture density, due to the main fault in the east, cannot be deduced (compare, for example, Locations 1 and 2, Fig. 4.3, with 200 m and 500 m distance to the fault, respectively, in Tabs. 4.1 and 4.2).

4.6.2 Fracture set analysis in a regional context

The observed main fracture orientations can be explained by the regional and local stress fields as represented by the faults in the quarry. Both NNW-SSE and W-E to WNW-ESE fault strikes are to be found throughout the Lower Saxony Basin (Fig. 4.2). An early Variscan deformation mechanism can be ruled out, as the study area is located north of the Variscan deformation zone. The NNW-SSE striking fractures are parallel to the NNW-SSE striking normal faults of the quarry and the region, and its onset may be either related to an extensional stress field prior to basin inversion in the burial phase in the Late Jurassic/Early Cretaceous (cf. Lohr et al., 2007; Sippel et al., 2009) or even earlier to the Rotliegend Graben structures, north of the Variscan Front (e.g. Doornebal and Stevenson, 2010). The onset of W-E to WNW-ESE striking faults, and associated fracture sets (WNW-ESE and WSW-ENE strike), may be related to an Late Jurassic/Early Cretaceous rotated extensional stress regime (Betz et al., 1987; Kley et al., 2008) or to NNE-SSW extension, as reported for the Keuper and Jurassic (e.g. Lohr et al., 2007; Kley et al., 2008); however, W-E to WNW-ESE striking normal faults may also be related to a local extensional stress field resulting from the folding of the uplifted block (e.g. Bissen, 2011), as a consequence of a N-S directed compression during basin inversion (Betz et al., 1987), which is characteristic of the southern part of the Central European Basin System (Kley and Voigt, 2008; Sippel et al., 2009). Engelder and Peacock (2001) and Glen et al. (2005) made similar observations of (local) extensional structures in an overall regional inversion/compressive regime at the southern margin of the Bristol Channel with comparable timing. WNW-ESE striking fractures were also reported for the unconformably-overlying Zechstein deposits (Voigt, 1960; Fiedler, 1984) and are locally sealed with subvertical quartz veins of equal strike around W-E striking normal faults (Wüstefeld et al., 2014; Wüstefeld et al., 2017a). This implies that at least fault-related fractures formed during or after the Mesozoic. The NNE-SSW striking fractures might be explained by a late rotation of the stress field, from a WNW-ESE to a WSW-ENE extensional direction, from Eocene times onwards (Kley and Voigt, 2008).

4.6.3 Evaluation of digital fracture detection

4.6.3.1 Limitations

General issues and constraints of digital outcrop models, and especially their acquisition, have been discussed in previous works (cf. references in the introduction), such as the impact of shading effects (e.g. Lato et al., 2010; Seers and Hodgetts, 2013). A careful visual assessment is indispensable, as all applied automated techniques may introduce artifact surfaces or false interpretations into the dataset (cf. Seers and Hodgetts, 2013, 2016a; Casini et al., 2016;). These might originate from non-geological processes, for example mining activities, but also from incorrect interpretations by the applied algorithm itself. Fracture sets measured on walls unmined for several years (southern wall) were compared to mined walls of different stages (Location 1, unmined *vs.* Locations 4 and 5, actively mined).

The primary assumption to detecting fracture discontinuity surfaces is their planarity (cf. Pollard and Aydin, 1988). Although providing a range of calibration options, automated detection can, for example, result in multiple fracture planes for highly irregular or intensely curved surfaces. In these cases, the accurate depiction of individual fractures cannot be assured, and manual intervention may be required. A thorough quality check of the automated analysis of the point data is therefore essential, and even a separated calibration for different fracture sets might be necessary to account for differences in exposed fracture styles. In the study presented here, the fracture curvature was found to be comparable for all fracture sets.

Similar to previous work, the outlined automated fracture detection approach exclusively detects fracture sets with well-exposed surfaces (e.g. Fernández, 2005; Sturzenegger and Stead, 2009; Larsen et al., 2010; García-Sellés et al., 2011; Seers and Hodgetts, 2013; Becker et al., 2014; Laux and Henk, 2015) and not fracture traces. This might result in a biased density estimation, if not all fracture surfaces are well-developed along the virtual scanline (e.g. Seers and Hodgetts, 2013); however, the possibility of analyzing several scanlines at the same location might reduce this bias, but it cannot be entirely ruled out. Moreover, fracture discontinuity surfaces are best-exposed subparallel to the quarry wall orientation, which reduces the aforementioned bias of the respective fracture sets. However, at the same time, this may result in a censoring of differently-oriented fracture orientations. Also, for the fracture spacing and density estimation based on the 1D scanlines, fracture spacing needs to be corrected in accordance with Terzaghi (1965). As the angle between fracture normal and outcrop wall orientation often exceeds 60°, in the case of the preferentially-detected fracture planes, it is likely that a bias of overestimated fracture densities may be introduced (cf. Bisdom et al., 2014). These potential bias sources demonstrate the necessity to investigate different quarry wall orientations, similar to conventionally-established manual field-based techniques, to reduce orientation-dependent biases respectively.

A suitable resolution of the acquired point cloud is needed for a reliable fracture detection, which must be represented by a sufficient amount of points. Small fracture discontinuity surfaces, therefore, might not be detected or may suffer inaccurate representations for a robust statistical distribution. Consequently, fracture set orientations exclusively or dominantly represented by the smallest fractures can be truncated or censored due to scanner resolution limitations, which might result in an underestimation (e.g. Ortega et al., 2006; Sturzenegger and Stead, 2009; Seers and Hodgetts, 2013; Bisdom et al., 2014).

4.6.3.2 Comparison with existing methods

Besides the advantage of conducting time-efficient fieldwork by digitizing the entire reservoir-scale outcrop analog (McCaffrey et al., 2005), the new workflow allows for quickly analyzing a large number of fractures over the entire exposed quarry dimension. This holds true for both the time needed for applying the outlined workflow and for computer processing time needed by the algorithm.

Manual fieldwork-derived structural surveys (e.g. Priest and Hudson, 1981; Mauldon et al., 2001), besides suffering disadvantages, such as inaccessibility of sampling sites (e.g. Casini et al., 2016), were found to be most precise with highest geological significance. The remote analysis, especially automated digital approaches, is not able to directly derive all essential geological information, which can be obtained by physical and direct work on the rocks, such as granulometry or mineralogy (e.g. Hodgetts, 2013; Casini et al., 2016); however, automated approaches result in statistically more robust datasets due to increased sample sizes acquired in short times (e.g. Fabuel-Perez et al., 2009; Gillespie et al., 2011; Hodgetts, 2013; Casini et al., 2016). Furthermore, the outlined digital scanline approach can avoid a human interpretation bias, which may affect manual field measurements, for example large-scale fracture discontinuity surfaces might be prone to preferred sampling.

Technically, the presented approach allows the direct investigation of acquired point clouds and does not need to rely on additional processing steps, such as mesh triangulation (e.g. Kemeny and Donovan, 2005; Mah et al., 2011; Vöge et al., 2013), which might introduce additional biases (Laux and Henk, 2015). Our outlined approach is based on classical PCA (e.g. Hotelling, 1933), which is conceptually similar to the method suggested by García-Sellés et al. (2011). Likewise to the work of Fernández (2005), the latter makes use of the orientation tensor analysis proposed by Woodcock (1977). Seers and Hodgetts (2016b) already noted that Woodcock's (1977) approach might be prone to numerical instability, if the center of mass is not equal to the origin of the coordinate system (cf. Seers and Hodgetts, 2016b). Furthermore, our presented algorithm integrates the PCA with a robust hierarchical clustering approach, enabling a direct and speedy recognition of the sought surface patches.

Although not being able to analyze discontinuity types without well-developed surfaces, such as veins or deformation bands, the outlined method enables the direct derivation of 3D orientation data of the detected fractures. This is not similarly possible by remote sensing techniques relying on the interpretation/trace mapping of 2D imagery (e.g. Barton et al., 1995; Le Garzic et al., 2011; Strijker et al., 2012; Hardebol and Bertotti, 2013; Bisdom et al., 2014). The application of fracture tracing techniques on 3D datasets (e.g. Seers and Hodgetts, 2013; Casini et al., 2016) may derive orientation data if the point sets do not have a collinear but a coplanar configuration, similar to exposed flat fracture surfaces, i.e. if a true 3D outcrop geometry is given; however, depending on the geometric properties, the actual (Seers and Hodgetts, 2016a) 3D orientation determination of lineament traces may still be erroneous (e.g. Seers and Hodgetts, 2016a, b). The probabilistic approach of Seers and Hodgetts (2016b) may improve tracing techniques by enhancing the accuracy of best-fit planes for structural lineaments. This method was applied to 3D fracture trace maps derived with calibrated 2D images (cf. Seers and Hodgetts, 2016a). Contrarily, the approach outlined in this work allows for the investigation also of quasi-2D outcrops, such as steep cliffs, and the direct derivation of orientation data. The presented concept of virtual scanlines has already been used in previous studies (e.g. Voyat et al., 2006; Seers and Hodgetts, 2013), but to the authors' knowledge, there have been no similar approaches, which directly and automatically assess point cloud data along virtual scanlines to quickly derive and visualize spatial heterogeneity of natural fractures.

4.6.4 T-LiDAR data as input for discrete fracture network modeling

The higher statistical significance, due to vast datasets with respect to spatial coverage, may result in more realistic input parameters for the modeling of a DFN (e.g. Wilson et al., 2011; Laux and Henk, 2015); however, it is noted that, for direct integration of spatially heterogeneous fracture data in reservoir models and in particular 3D extrapolation, an established, complex 3D model is required, implying that obtained high-resolution data cannot be integrated straightforwardly into low-resolution simplified models.

The drawback of the approach is a potential bias when fractures may not be represented as well-developed surfaces or the inability to investigate other discontinuity types, exclusively represented by lineament traces, such as veins or deformation bands; however, if at all, the latter discontinuity types need to be modeled separately because of substantial differences in hydraulic properties, when compared to fractures. A combination of the outlined approach with tracing techniques (e.g. Casini et al., 2016; Seers and Hodgetts, 2016a) may be a promising future development, which might benefit from the advantages of both principles of fracture detection (cf. previous section).

Data needs to be integrated into detailed reservoir models to be able to model complex discrete natural fracture networks. For the construction of complex 3D models, the integration of robust facies models along with detailed structural framework models is of great significance. Rarity et al. (2014), for example, presented promising workflows to integrate t-LiDAR data into reservoir modeling software, with a focus on geocellular and stochastic facies modeling, enabling the representation of the 3D complexity of sedimentary systems. The incorporation of new approaches is necessary to progress towards full data integration of digital outcrop analog data into modern exploration and production industry modeling workflows at an optimal grid resolution. Nevertheless, classical field-based geology is indispensable in order to validate automated fracture detection and cannot be entirely substituted by digital geological approaches.

4.7 Conclusions

The presented workflow integrates high-resolution digital natural fracture network data, derived from terrestrial laser scanning, into modern petroleum industry workflows. The algorithm devised enables the rapid automated detection of fracture surfaces along defined virtual scanlines and was successfully tested with synthetic virtual scanline data and manual outcrop data. The automated detection of fracture sets is thus a valid option for fracture analysis over a100s of meters scale.

A NNW-SSE fracture set is present in the entire quarry, while a WNW-ESE striking set is dominant in the west of the quarry. In the east, the latter is subordinate and WSW-ENE striking fractures are prevalent. The analysis shows that fractures rotate laterally over a 100s of meters scale across the study area, with a W-E strike changing towards a WSW-ENE strike close to the main NNW-SSE striking fault in the east. This indicates the effects of fracture rotation due to stress shadows around large faults.

The fracture density across same facies units varies between 1 m^{-1} and 7 m^{-1} with an average of 3.4 to 3.9 m^{-1}. Zones of higher fracture densities are 10 to 20 m wide, and can be traced laterally along virtual scanlines from terrestrial laser scanning data. These fracture corridors are not related to the minor W-E striking normal faults crossing the virtual scanlines.

Fracture densities and orientations are lithology dependent. The density of a single NNE-SSW set is higher (0.9 m^{-1}), and the strike of the main fracture set is approximately 5 to 10° rotated clockwise, in a fine- to medium-grained lacustrine sandstone than in the underlying scanlines along coarse- to medium-grained fluvial sandstones (0.55 m^{-1}). This likely highlights the effect of differences in elastic properties of the different lithotypes.

Both the t-LiDAR data and the fracture statistics generated from automated scanline analyses of t-LiDAR data were successfully integrated into standard exploration and production industry modeling software with respect to orientation and spatial distribution characteristics.

This automated scanline approach may improve the characterization of fracture networks in field-scale outcrop analogs. The ability to describe and display the spatial variation of fracture statistics is essential to understanding the complexity of the fracture network and its integration into DFN models. The approach opens up a variety of different applications, such as the assessment of the impact of various borehole orientations on the characterization of fracture density distributions. This study may contribute to an improved application of outcrop analog data to naturally fractured reservoirs in the subsurface, reducing uncertainties in the characterization of this type of reservoir at depth.

5 Conclusions and outlook

5.1 Résumé

This study focused on the reservoir characteristics of Upper Carboniferous tight gas sandstones. The reservoir outcrop analog at the southwestern rim of the Lower Saxony Basin was investigated by integrating structural, stratigraphic and diagenetic data. The study area is located at the intersection of two normal fault strike directions and reflects the structure of some of the producing Upper Carboniferous tight gas fields in the North German Basin. The sedimentology and the dimension of the quarry, as well as sedimentary bodies, match and reflect subsurface reservoirs.

The diagenetic characterization reveals an early loss of primary porosity based on the formation of a pseudomatrix as a result of compaction in the eogenetic realm. The present day porosity in the outcrop analog is laterally consistent in single horizons and ranges between 4.5 to 8.7 %. Average porosities match with NW German tight gas reservoirs at depth. However, the diagenetic reconstruction uncovered that porosity of the outcrop analog was considerably lower during burial. The dissolution of carbonates and unstable alumosilicates mainly contributed to present day macroporosity and can be predominantly attributed to telodiagenesis. Leaching processes were favored in fault damage zones of W-E striking faults because of increased fluid flow. This led to enhanced matrix porosities of up to 26.3 % and permeabilities up to 105 mD. However, a telogenetic origin cannot be ruled out and the increased reservoir properties in fault corridors cannot be accounted for by reservoir-relevant processes based on the available data.

The present day permeability of the outcrop analog is one to two orders of magnitude lower than the compared subsurface reservoir. The low permeability is the result of a thermal overprint, which is represented by intense mesogenetic quartz cementation and illitization. Mesogenetic meshwork illite was impregnated with bitumen as a consequence of hydrocarbon charging. In the ongoing burial, authigenic quartz cemented the meshwork illite, well-reflecting the thermal overprint of the tight gas siliciclastics. Based on the correlation of petrographic and petrophysical data, no reservoir quality controlling trends can be established. The absence of potential facies-related trends may imply that these were overprinted by high-temperature diagenesis. As such, in the case of the study site, the temperature overprint was the controlling factor of reservoir properties during mesodiagenesis.

The characterization of the thermal event, based on chlorite thermometry, fluid inclusion microthermometry and vitrinite reflectance measurements, infers that the investigated rocks were exposed to temperatures of approximately 250 – 300°C. This local temperature increase was approximately 90 – 120°C higher when compared to similar stratigraphic positions in the region. At the study site, it was the result of circulating hydrothermal fluids along the fault damage zone of a large NNW-SSE striking normal fault with a displacement of up to 600 m. The lateral heating of the formation of up to one km away from the fault is evidenced by vitrinite reflectance data of the investigated meta-anthracite coals as well as the temperature-related diagenetic overprint of the entire study site. K-Ar dating and stable isotope data of fluid inclusions suggest that highest coalification was already reached during Mid to Late Jurassic rifting (162 Ma, < 2 µm fraction) before basin inversion (90 – 100 Ma). The data show that coalification was not subject to significant change during ongoing burial history, which is further constrained by the biaxial reflectance anisotropy of meta-anthracite coals.

The rapid improvement of technologies employed in digital field geology allows for the acquisition of higher levels of accuracy when characterizing field analogs. The outlined workflow supports the integration of high-resolution digital field mapping data based on terrestrial laserscanning (t-LiDAR) into modern industry workflows for characterizing natural fracture networks. Thereby, the presented algorithm allows for automated fracture surface detection along defined scanlines, which was successfully tested with synthetic scanline data and validated against traditional manual scanline data. The analysis of the study site revealed that fractures rotate laterally over a 100s meter scale across the study area, indicating fracture rotation from stress shadows around large faults. On a 10s of meters scale, zones of higher fracture densities were detected along the investigated scanlines, which cannot be related to faulting. Fracture densities and orientations are lithology dependent with different distributions between coarse- to medium-grained fluvial sandstones and fine- to medium grained lacustrine sandstones, likely highlighting the effect of differences in elastic properties of the different lithotypes

5.2 General implications

The results of this work demonstrate both the transferability and limits of outcrop analog studies. The Piesberg quarry forms a suitable analog with respect to sedimentological, stratigraphic and structural inventory as well as detrital and authigenic composition within the sedimentary bodies. The relative porosity and permeability trends are comparable and maybe also their spatial heterogeneity. However, subsurface reservoirs at depth generally lack telogenetic influences, altering reservoir properties at the surface. The investigated outcrop is a reservoir analog with respect to fault-induced thermal anisotropies, and their implications for maturity and associated diagenesis, but not for a closed-system tight gas reservoir rock. Faults are thus expected to have considerable impact on reservoir qualities. At first sight, the outcrop analog has apparently similar average porosity values compared to the subsurface reservoir. However, the outcrop does not reflect absolute porosity and permeability values because the diagenetic evolution is different and the result of the thermal and telogenetic overprint. The results demonstrate that a fault with approximately 600 m offset can increase maturity over a kilometer scale, forming "sweet spots" in otherwise immature regions. In the presented case study, the NNW-striking fault and related thermal anomaly pose a risk to reservoir quality, resulting in overmature source rocks and associated diagenetic cementation. The latter deteriorates reservoir properties with respect to porosity and permeability.

The study has also shown that peak temperatures cannot always be associated with peak subsidence during burial. Fault-related local thermal anomalies during burial can be higher than peak subsidence temperatures. Such fault-induced temperature anomalies thus cannot be excluded in the modeling of petroleum systems and reservoir quality prediction in tight sandstones. The neglect of such local phenomena affecting reservoir properties laterally over more than one kilometer away may result, for instance, in overestimates of burial depths and misinterpretations of recoverable resources.

The outlined digital field geology approach based on terrestrial laser scanning (t-LiDAR) may improve fracture network characterization in field-scale outcrop analogs. It permits the assessing of large datasets at different scales and integrating them into 3D digital outcrop models. This fosters understanding the complexity and spatial heterogeneity of natural fracture networks and is essential for establishing DFN models. The workflow devised may contribute to an improved application of outcrop analog data to naturally fractured reservoirs in the subsurface and may thus reduce uncertainty in the characterization of analogous reservoir types. Fracture corridors can be detected, which may host "sweet spots" in tight sandstones banned from fracking.

5.3 Outlook

This work showed that fault-bound circulating hydrothermal fluids, laterally heating up formation, are the most likely scenario for the local thermal anomaly. The results should be compared with subsurface data of producing fields and/or exploration wells. Such data may validate the outlined scenario and provide further insights into fault-bound thermal anomalies at comparable structural positions within the region. A 3D numerical modeling approach may further unravel the validity and plausibility of the proposed model of the thermal overprint, with special emphasis on the convective heat/mass transfer mechanism in the fault damage zone. It may be of interest to establish burial scenarios based on petroleum system modeling, which incorporate the novel results of this study. Such investigations should primarily focus on the proposed timing of the thermal overprint and estimated burial depth as presented in Chapter 3. Hypothesizing the absence of previously postulated intrusive bodies and assuming the smaller maximum burial depth would imply a reevaluation of the region, involving, for example, a new assessment of the prospectivity with respect to the presence of Paleozoic and Mesozoic gas fields (e.g. Brink, 2013).

Based on the available data, the corridors of increased reservoir properties could not be associated with a subsurface-relevant dissolution process. However, such corridors were shown to be related to W-E striking normal faults. A second fluid flow event at times of basin inversion was verified and is amongst others manifested by ore mineralization associated with W-E striking normal faults. Similar ore mineralization is frequently reported to be linked with the structural inventory in the region (e.g. Klassen, 1984; Röhrs, 1992). Future studies may concentrate on a genetic relationship between reservoir properties and this late fluid flow event, which is likely associated with the structural inventory resulting from basin inversion. The integration of outcrop and subsurface data of several study locations may be a key to unravelling these genetic relationships and possibly associated increased dissolution potential. Therefore, ore mineralization could be used as proxy for dissolution events enhancing reservoir properties around faults.

The presented automated workflow for terrestrial laser scanning (t-LiDAR) has the potential of contributing significantly to the characterization of fracture networks of outcrop analogs. However, the current stage is a promising prototype workflow. Besides the great possibilities with this approach, it still requires further elaboration. This includes the study, verification and application of the outlined approach on a variety of different outcrop analogs with various structural as well as sedimentological styles and scales. The workflow does not yet allow the simultaneous derivation of further fracture characteristics, such as fracture surface area, at the current developmental stage. The presented algorithm is not restricted to 1D virtual scanlines and enables the analysis of entire quarry walls, which bears further data, such as, for instance, the fracture surface area. Future development should incorporate techniques to derive fracture persistence, trace length and additional statistical relevant parameters, which are included in the vast t-LIDAR dataset. Data needs to be integrated into detailed reservoir models in order to be able to model complex discrete natural fracture networks based on the available high-resolution spatial fracture data as well as to evaluate the same and to eventually conduct upscaling processes as well as flow simulations for a detailed advanced evaluation. However, the author likes to underscore that classical field geology is indispensable and cannot be entirely substituted by digital geological approaches.

References

Adriasola-Muñoz, Y., 2006, The thermal history of the western Lower Saxony Basin, Germany. PhD thesis, RWTH Aachen University, Aachen, 152 pp.

Adriasola-Muñoz, Y., Littke, R., Brix, M.R., 2007. Fluid systems and basin evolution of the western Lower Saxony Basin, Germany. Geofluids 7, 335-355.

Ajdukiewicz, J.M., Lander, R.H., 2010. Sandstone reservoir quality prediction: The state of the art. AAPG Bulletin 94, 1083-1091.

Albrecht, D., Reitenbach, V., 2015. Investigations on fluid transport properties in the North-German Rotliegend tight gas sandstones and applications. Environmental Earth Sciences 73, 5791-5799.

Alfarhan, M., White, L., Tuck, D., Aiken, C., 2008. Laser rangefinders and ArcGIS combined with three-dimensional photorealistic modeling for mapping outcrops in the Slick Hills, Oklahoma. Geosphere 4, 576-587.

Allen, P.A., Allen, J.R., 2005. Basin Analysis: Principles and applications, 2nd ed. Blackwell Scientific Publications, Oxford, United Kingdom.

Amann-Hildenbrand, A., Dietrichs, J.P., Krooss, B.M., 2016. Effective gas permeability of Tight Gas Sandstones as a function of capillary pressure – a non-steady-state approach. Geofluids 16, 367-383.

Andrews, J.R., Day, J., Marshall, J.E.A., 1996. A thermal anomaly associated with the Rusey Fault and its implications for fluid movements. Proceedings of the Ussher Society 9, 68-71.

Angenheister, G.E., 1982. Landolt-Börnstein: Zahlenwerte und Funktionen aus Naturwissenschaft und Technik. Neue Serie, Gruppe 5: Geophysik und Weltraumforschung, Band 1: Physikalische Eigenschaften der Gesteine. Springer, Berlin, Germany.

Bachmann, G.H., Grosse, S., 1989. Struktur und Entstehung des Norddeutschen Beckens - geologische und geophysikalische Interpretation einer verbesserten Bouguer-Schwerekarte. Niedersächsische Akademie der Geowissenschaften Veröffentlichungen 2, 23-47.

Bahrami, H., Rezaee, M.R., Nazhat, D.H., Ostojic, J., Clennell, M.B., Jamili, A., 2011. Effect of Water Blocking Damage on Flow Efficiency and Productivity in Tight Gas Reservoirs, SPE Production and Operations Symposium. Society of Petroleum Engineers.

Bahrami, H., Rezaee, R., Hossain, M., 2012. Characterizing natural fractures productivity in tight gas reservoirs. Journal of Petroleum Exploration and Production Technology 2, 107-115.

Bakker, R.J., 2003. Package FLUIDS 1. Computer programs for analysis of fluid inclusion data and for modelling bulk fluid properties. Chemical Geology 194, 3-23.

Baldschuhn, R., Binot, F., Fleig, S., Kockel, F., 2001. Geotektonischer Atlas von Nordwest-Deutschland und dem deutschen Nordsee-Sektor. Geologisches Jahrbuch (A) 153, 3-95.

Baldschuhn, R., Kockel, F., 1999. Das Osning-Lineament am Südrand des Niedersachsen-Beckens. Zeitschrift der deutschen geologischen Gesellschaft 150, 673-695.

Bartenstein, H., Teichmüller, M., Teichmüller, R., 1971. Die Umwandlung der organischen Substanz im Dach des Bramscher Massivs. Fortschritte in der Geologie von Rheinland und Westfalen 18, 501-538.

Barthélémy, J.-F., Guiton, M.L.E., Daniel, J.-M., 2009. Estimates of fracture density and uncertainties from well data. International Journal of Rock Mechanics and Mining Sciences 46, 590-603.

Barton, C.A., Zoback, M.D., Moos, D., 1995. Fluid flow along potentially active faults in crystalline rock. Geology 23, 683-686.

Bässler, R., Hoyer, P., 1971. Das Karbon von Ibbenbüren, am Hüggel und Piesberg. Fortschritte in der Geologie von Rheinland und Westfalen 19, 79-82.

Bazin, B., Bekri, S., Vizika, O., Herzhaft, B., Aubry, E., 2010. Fracturing in tight gas reservoirs: Application of special-core-analysis methods to investigate formation-damage mechanisms. SPE Journal 15, 975-982.

Becker, I., Wüstefeld, P., Koehrer, B., Felder, M., Hilgers, C., 2017. Porosity and permeability variations in a tight gas sandstone reservoir analogue, Westphalian D, Lower Saxony Basin, NW Germany: influence of depositional settings and diagenesis. Journal of Petroleum Geology 40, 363-389.

Becker, S., Hilgers, C., Kukla, P.A., Urai, J.L., 2011. Crack-seal microstructure evolution in bi-mineralic quartz-chlorite veins in shales and siltstones from the RWTH-1 Well, Aachen, Germany. Journal of Structural Geology 33, 676-689.

Becker, S., Nguyen, H.T., Nollet, S., Fernandez-Steeger, T.M., Laux, D., Hilgers, C., 2014. Methods to analyse fracture orientation patterns in a Lower Carboniferous carbonate reservoir analogue in the Voreifel, Germany. Zeitschrift der Deutschen Gesellschaft für Geowissenschaften 165, 319-330.

Behr, H.J., Horn, E.E., Frentzel-Beyme, K., Reutel, C., 1987. Fluid inclusion characteristics of the Variscan and post-Variscan mineralizing fluids in the Federal Republic of Germany. Chemical Geology 61, 273-285.

Bellian, J.A., Kerans, C., Jennette, D.C., 2005. Digital Outcrop Models: Applications of Terrestrial Scanning Lidar Technology in Stratigraphic Modeling. Journal of Sedimentary Research 75, 166-176.

Bergmann, J., Kleeberg, R., 1998. Rietveld Analysis of Disordered Layer Silicates. Material Science Forum 278, 300-305.

Bertier, P., Swennen, R., Lagrou, D., Laenen, B.E.N., Kemps, R., 2008. Palaeo-climate controlled diagenesis of the Westphalian C & D fluvial sandstones in the Campine Basin (north-east Belgium). Sedimentology 55, 1375-1417.

Besly, B.M., 1998. Carboniferous, in: Glennie, K.W. (Ed.), Petroleum Geology of the North Sea: Basic Concepts and Recent Advances, 4th ed. Blackwell Science Ltd., Oxford, United Kingdom, pp. 104-136.

Betz, D., Führer, F., Greiner, G., Plein, E., 1987. Evolution of the Lower Saxony Basin. Tectonophysics 137, 127-170.

Betzer, H.J., Dassel, W., Drozdzewski, G., Farrenschon, J., Gawlik, A., Heuser, H., Juch, D., Ribbert, K.H., Skupin, K., Klassen, H., Bérenger, D., 2003. Geologie im Weser- und Osnabrücker Bergland. Geologischer Dienst Nordrhein-Westfalen, Krefeld, Federal Republic of Fermany.

BGR, 2016. Energiestudie 2016. Reserven, Ressourcen und Verfügbarkeit von Energierohstoffen, Hannover, Germany, p. 180.

Bing Maps, 2015. Piesberg, Osnabrück, Federal Republic of Germany.

Bisdom, K., Gauthier, B.D.M., Bertotti, G., Hardebol, N.J., 2014. Calibrating discrete fracture-network models with a carbonate three-dimensional outcrop fracture network: Implications for naturally fractured reservoir modeling. AAPG Bulletin 98, 1351-1376.

Bissen, R., 2011, Charakterisierung und Modellierung von Kluftnetzwerken am Beispiel des Piesberges. PhD thesis, Albert-Ludwigs-Universität, Freiburg, Germany, 216 pp.

Bjørlykke, K., Egeberg, P.K., 1993. Quartz cementation in sedimentary basins. AAPG Bulletin 77, 1538-1548.

Bjørlykke, K., Jahren, J., 2012. Open or closed geochemical systems during diagenesis in sedimentary basins: Constraints on mass transfer during diagenesis and the prediction of porosity in sandstone and carbonate reservoirs. AAPG Bulletin 96, 2193-2214.

Bjørlykke, K., Mo, A., Palm, E., 1988. Modelling of thermal convection in sedimentary basins and its relevance to diagenetic reactions. Marine and Petroleum Geology 5, 338-351.

Boles, J.R., 1978. Active ankerite cementation in the subsurface Eocene of Southwest Texas. Contributions to Mineralogy and Petrology 68, 13-22.

Bonnet, E., Bour, O., Odling, N.E., Davy, P., Main, I., Cowie, P., Berkowitz, B., 2001. Scaling of fracture systems in geological media. Reviews of Geophysics 39, 347-383.

Bottinga, Y., 1969. Calculated fractionation factors for carbon and hydrogen isotope exchange in the system calcite-carbon dioxide-graphite-methane-hydrogen-water vapor. Geochimica et Cosmochimica Acta 33, 49-64.

Bourbiaux, B., 2010. Fractured Reservoir Simulation: a Challenging and Rewarding Issue. Oil and Gas Science and Technology – Revue de l'Institut Francais du Pétrole 65, 227-238.

Bourdelle, F., Parra, T., Chopin, C., Beyssac, O., 2013. A new chlorite geothermometer for diagenetic to low grade metamorphic conditions. Contributions to Mineralogy and Petrology 165, 723-735.

Bourne, S.J., 2003. Contrast of elastic properties between rock layers as a mechanism for the initiation and orientation of tensile failure under uniform remote compression. Journal of Geophysical Research 108.

Bourne, S.J., Willemse, E.J.M., 2001. Elastic stress control on the pattern of tensile fracturing around a small fault network at Nash Point, UK. Journal of Structural Geology 23, 1753-1770.

Bowitz, J., Ehling, A., 2008. Non-destructive infrared analyses: a method for provenance analyses of sandstones. Environmental Geology 56, 623-630.

Breyer, F., 1971. Geophysikalische und geologische Beiträge zur oberflächennahen Tektonik im Dach des Bramscher Massivs. Fortschritte in der Geologie von Rheinland und Westfalen 18, 354-385.

Brink, H.J., 2013. Die Intrusion von Bramsche - ein Irrtum im invertierten Niedersächsischen Becken? Zeitschrift der Deutschen Gesellschaft für Geowissenschaften 164, 33-48.

Brink, H.J., Dürschner, H., Trappe, H., 1992. Some aspects of the late and post-Variscan development of the Northwestern German Basin. Tectonophysics 207, 65-95.

Brockamp, B., 1967. Kurzbericht über die im Gebiet um Osnabrück durchgeführten seismischen Arbeiten des Institutes für reine und angewandte Geophysik der Universitat Münster. Veröffentlichungen der Deutschen Geodätischen Kommission (B) 153, 1-12.

Bruhn, R.L., Parry, W.T., Yonkee, W.A., Thompson, T., 1994. Fracturing and hydrothermal alteration in normal fault zones. Pure and Applied Geophysics 142, 609-644.

Bruns, B., di Primio, R., Berner, U., Littke, R., 2013. Petroleum system evolution in the inverted Lower Saxony Basin, northwest Germany: a 3D basin modeling study. Geofluids 13, 246-271.

Bruns, B., Littke, R., 2015. Lithological dependency and anisotropy of vitrinite reflectance in high rank sedimentary rocks of the Ibbenbüren area, NW-Germany: Implications for the tectonic and thermal evolution of the Lower Saxony Basin. International Journal of Coal Geology 137, 124-135.

Bryant, I., Carr, D., Cirilli, P., Drinkwater, N., McCormick, D., Tilke, P., Thurmond, J., 2000. Use of 3D digital analogues as templates in reservoir modelling. Petroleum Geoscience 6, 195-201.

Buckley, S.J., Enge, H.D., Carlsson, C., Howell, J.A., 2010. Terrestrial laser scanning for use in virtual outcrop geology. The Photogrammetric Record 25, 225-239.

Buckley, S.J., Howell, J.A., Enge, H.D., Kurz, T.H., 2008. Terrestrial laser scanning in geology: data acquisition, processing and accuracy considerations. Journal of the Geological Society of London 165, 625-638.

Burton, D., Dunlap, D.B., Wood, L.J., Flaig, P.P., 2011. Lidar intensity as a remote sensor of rock properties. Journal of Sedimentary Research 81, 339-347.

Casini, G., Hunt, D.W., Monsen, E., Bounaim, A., 2016. Fracture characterization and modeling from virtual outcrops. AAPG Bulletin 100, 41-61.

Cawood, A.J., Bond, C.E., Howell, J.A., Butler, R.W.H., Totake, Y., 2017. LiDAR, UAV or compass-clinometer? Accuracy, coverage and the effects on structural models. Journal of Structural Geology 98, 67-82.

Cherubini, Y., Cacace, M., Bloecher, G., Scheck-Wenderoth, M., 2013. Impact of single inclined faults on the fluid flow and heat transport: results from 3-D finite element simulations. Environmental Earth Sciences 70, 3603-3618.

Clark, R.N., Roush, T.L., 1984. Reflectance spectroscopy: Quantitative analysis techniques for remote sensing applications. Journal of Geophysical Research: Solid Earth 89, 6329-6340.

Cowan, G., 1989. Diagenesis of Upper Carboniferous sandstones: southern North Sea Basin. Geological Society, London, Special Publications 41, 57-73.

Cowie, P.A., Scholz, C.H., 1992. Displacement-length scaling relationship for faults; data synthesis and discussion. Journal of Structural Geology 14, 1149-1156.

Curewitz, D., Karson, J.A., 1997. Structural settings of hydrothermal outflow: Fracture permeability maintained by fault propagation and interaction. Journal of Volcanology and Geothermal Research 79, 149-168.

Curtis, C.D., 1985. Clay mineral precipitation and transformation during burial diagenesis. Philosophical Transactions of the Royal Society of London, Series A: Mathematical and Physical Sciences 315, 91-105.

Curtis, C.D., Hughes, C.R., Whiteman, J.A., Whittle, C.K., 1985. Compositional variation within some sedimentary chlorites and some comments on their origin. Mineralogical Magazine 49, 375-386.

Dai, J., Yan, S., Dai, C., Wang, D., 1996. Geochemistry and accumulation of carbon dioxide gases in China. AAPG Bulletin 80, 1615-1626.

David, C., Wong, T.-F., Zhu, W., Zhang, J., 1994. Laboratory measurement of compaction-induced permeability change in porous rocks: Implications for the generation and maintenance of pore pressure excess in the crust. Pure and Applied Geophysics 143, 425-456.

David, F., 1987. Sandkörper in fluviatilen Sandsteinen des Unteren Westfal D (Oberkarbon) am Piesberg bei Osnabrück. Facies 17, 51-58.

David, F., 1990. Sedimentologie und Beckenanalyse im Westfal C und D des nordwestdeutschen Oberkarbons, DGMK-Bericht 384-3, Sedimentologie des Oberkarbons. Deutsche Gesellschaft für Mineralölwissenschaft und Kohlechemie e.V., Hamburg, Germany, p. 267.

de Caritat, P., Hutcheon, I., Walshe, J.L., 1993. Chlorite geothermometry: a review. Clays and Clay Minerals 41, 219-239.

Deer, W.A., Howie, R.A., Zussman, J., 1992. An introduction to the rock-forming minerals. Longman, London, United Kingdom.

Dershowitz, W., 1985, Rock joint systems. PhD thesis, Massachusetts Institute of Technology, Cambridge, Massachusetts, USA.

Dershowitz, W.S., Herda, H.H., 1992. Interpretation of fracture spacing and intensity. Proceedings - Symposium on Rock Mechanics 33, 757-766.

Doornebal, J.C., Stevenson, A.G., 2010. Petroleum Geological Atlas of the Southern Permian Basin Area. EAGE Publications b.v., Houten, Netherlands.

Drozdzewski, G., 1985. Tiefentektonik der Ibbenbürener Karbon-Scholle, Beiträge zur Tiefentektonik westdeutscher Steinkohlenlagerstätten. Geologisches Landesamt Nordrhein-Westfalen Krefeld, pp. 189-216.

Drozdzewski, G., 1988. Die Wurzel der Osning-Überschiebung und der Mechanismus herzynischer Inversionsstörungen in Mitteleuropa. Geologische Rundschau 77, 127-141.

Drozdzewski, G., 1998. Tektonische Analyse des Permokarbon zur Prognose des Kluftpotentials im Raum zwischen Hannover und der Weserscholle. - Tektonische Studie im Auftrag der Mobil Erdgas-Erdöl GmbH und der Preussag Energie Gmbh (unpublished), Krefeld, p. 77.

Drozdzewski, G., Henscheid, S., Hoth, P., Juch, D., Littke, R., Vieth, A., Wrede, V., 2009. The pre-Permian of NW-Germany - structure and coalification map. Zeitschrift der Deutschen Gesellschaft für Geowissenschaften 160, 159-172.

Drozdzewski, G., Ribbert, K.H., 2003. Erdgeschichte, in: Geologischer Dienst Nordrhein-Westfalen (Ed.), Geologie im Weser- und Osnabrücker Bergland. Geologischer Dienst Nordrhein-Westfalen, Krefeld, Germany, p. 31.

Drozdzewski, G., Schäfer, A., Brix, M.R., 2008. Field trip PRE2: Late Carboniferous clastic sequences of the southern Ruhr Basin, (Germany). Exkursionsführer und

Veröffentlichungen der Deutschen Gesellschaft fuer Geowissenschaften 237, 5-20.

Duschl, F., van den Kerkhof, A., Sosa, G., Leiss, B., Wiegand, B., Vollbrecht, A., Sauter, M., 2016. Fluid inclusion and microfabric studies on Zechstein carbonates (Ca2) and related fracture mineralizations – New insights on gas migration in the Lower Saxony Basin (Germany). Marine and Petroleum Geology 77, 300-322.

Dutton, S.P., Laubach, S.E., Tye, R.S., Baumgardner, R.W., Jr., Herrington, K.L., 1991. Geologic characterization of low-permeability gas reservoirs, Travis Peak Formation, East Texas. University of Texas at Austin, Bureau of Economic Geology, Austin, TX, United States.

Ehrenberg, S.N., 1995. Measuring sandstone compaction from modal analyses of thin sections: how to do it and what the results mean. Journal of Sedimentary Research 65, 369-379.

Enge, H.D., Howell, J.A., Buckley, S.J., 2010. Quantifying clinothem geometry in a forced-regressive river-dominated delta, Panther Tongue Member, Utah, USA. Sedimentology 57, 1750-1770.

Enge, H.v.D., Buckley, S.J., Rotevatn, A., Howell, J.A., 2007. From outcrop to reservoir simulation model: Workflow and procedures. Geosphere 3, 469-490.

Engelder, T., Peacock, D.C.P., 2001. Joint development normal to regional compression during flexural-flow folding: the Lilstock buttress anticline, Somerset, England. Journal of Structural Geology 23, 259-277.

Essene, E.J., Peacor, D.R., 1995. Clay mineral thermometry - a critical perspective. Clays and Clay Minerals 43, 540-553.

Fabuel-Perez, I., Hodgetts, D., Redfern, J., 2009. A new approach for outcrop characterization and geostatistical analysis of a low-sinuosity fluvial-dominated succession using digital outcrop models: Upper Triassic Oukaimeden Sandstone Formation, central High Atlas, Morocco. AAPG Bulletin 93, 795-827.

Fabuel-Perez, I., Hodgetts, D., Redfern, J., 2010. Integration of digital outcrop models (DOMs) and high resolution sedimentology – workflow and implications for geological modelling: Oukaimeden Sandstone Formation, High Atlas (Morocco). Petroleum Geoscience 16, 133-154.

Faulkner, D.R., Jackson, C.A.L., Lunn, R.J., Schlische, R.W., Shipton, Z.K., Wibberley, C.A.J., Withjack, M.O., 2010. A review of recent developments concerning the structure, mechanics and fluid flow properties of fault zones. Journal of Structural Geology 32, 1557-1575.

Faulkner, D.R., Mitchell, T.M., Jensen, E., Cembrano, J., 2011. Scaling of fault damage zones with displacement and the implications for fault growth processes. Journal of Geophysical Research 116, B05403.

Fernández, O., 2005. Obtaining a best fitting plane through 3D georeferenced data. Journal of Structural Geology 27, 855-858.

Fernández, O., Muñoz, J.A., Arbués, P., Falivene, O., Marzo, M., 2004. Three-dimensional reconstruction of geological surfaces: An example of growth strata and turbidite systems from the Ainsa basin (Pyrenees, Spain). AAPG Bulletin 88, 1049-1068.

Ferrero, A.M., Forlani, G., Roncella, R., Voyat, H.I., 2008. Advanced Geostructural Survey Methods Applied to Rock Mass Characterization. Rock Mechanics and Rock Engineering 42, 631-665.

Fiedler, K., 1984. Tektonik, in: Klassen, H. (Ed.), Geologie des Osnabrücker Berglandes. Naturwissenschaftliches Museum Osnabrück, Germany, pp. 519-567.

Finkbeiner, T., Barton, C.A., Zoback, M.D., 1997. Relationships among in-situ stress, fractures and faults, and fluid flow: Monterey formation, Santa Maria Basin, California. American Association of Petroleum Geologists Bulletin 81, 1975-1999.

Fleming, C.G., Couples, G.D., Haszeldine, R.S., 1998. Thermal effects of fluid flow in steep fault zones. Geological Society, London, Special Publications 147, 217-229.

Florez-Nino, J.-M., Aydin, A., Mavko, G., Antonellini, M., Ayaviri, A., 2005. Fault and fracture systems in a fold and thrust belt: an example from Bolivia. AAPG Bulletin 89, 471-493.

Fossen, H., Hesthammer, J., 1997. Geometric analysis and scaling relations of deformation bands in porous sandstone. Journal of Structural Geology 19, 1479-1493.

Franke, D., 1995. The North Variscan Foreland, in: Dallmeyer, R.D., Franke, W., Weber, K. (Eds.), Pre-Permian Geology of Central and Eastern Europe. Springer-Verlag, Berlin, Germany, pp. 554-566.

Frings, K., Lutz, R., de Wall, H., Warr, L.N., 2004. Coalification history of the Stephanian Ciñera-Matallana pull-apart basin, NW Spain: Combining anisotropy of vitrinite reflectance and thermal modelling. International Journal of Earth Sciences 93, 92-106.

Füchtbauer, H., 1974. Zur Diagenese fluviatiler Sandsteine. Geologische Rundschau 63, 904-925.

Füchtbauer, H., 1988. Sedimente und Sedimentgesteine. Sediment-Petrologie, Teil II. Schweizerbart, Stuttgart, Federal Republic of Germany.

García-Sellés, D., Falivene, O., Arbués, P., Gratacos, O., Tavani, S., Muñoz, J.A., 2011. Supervised identification and reconstruction of near-planar geological surfaces from terrestrial laser scanning. Computers & Geosciences 37, 1584-1594.

Gaupp, R., Matter, A., Platt, J., Ramseyer, K., Walzebuck, J., 1993. Diagenesis and fluid evolution of deeply buried Permian (Rotliegende) gas reservoirs, Northwest Germany. AAPG Bulletin 77, 1111-1128.

Gaupp, R., Möller, P., Lüders, V., di Primio, R., Littke, R., 2008. Fluids in sedimentary basins: an overview, in: Littke, R., Bayer, U., Gajewski, D., Nelskamp, S. (Eds.), Dynamics of complex intracontinental basins. Springer, Berlin, Federal Republic of Germany, pp. 347-365.

Gaupp, R., Okkerman, J.A., 2011. Diagenesis and reservoir quality of Rotliegend sandstones in the northern Netherlands - a review. Special Publication - Society for Sedimentary Geology 98, 193-226.

Gautier, D.L., 2003. Carboniferous-Rotliegend total petroleum system: description and assessment results, U. S. Geological Survey Bulletin. U. S. Geological Survey, Reston, VA, United States.

Ghanizadeh, A., Gasparik, M., Amann-Hildenbrand, A., Gensterblum, Y., Krooss, B.M., 2013. Lithological Controls on Matrix Permeability of organic-rich Shales: An Experimental Study. Energy Procedia 40, 127-136.

Ghanizadeh, A., Gasparik, M., Amann-Hildenbrand, A., Gensterblum, Y., Krooss, B.M., 2014. Experimental study of fluid transport processes in the matrix system of the

European organic-rich shales: I. Scandinavian Alum Shale. Marine and Petroleum Geology 51, 79-99.

Ghosh, K., Mitra, S., 2009. Structural controls of fracture orientations, intensity, and connectivity, Teton Anticline, Sawtooth Range, Montana. AAPG Bulletin 93, 995-1014.

Gigli, G., Casagli, N., 2011. Semi-automatic extraction of rock mass structural data from high resolution LIDAR point clouds. International Journal of Rock Mechanics and Mining Sciences 48, 187-198.

Gillespie, P., Monsen, E., Maerten, L., Hunt, D., Thurmond, J., Tuck, D., 2011. Fractures in Carbonates: From Digital Outcrops to Mechanical Models. Concepts in Sedimentology and Paleontology 10, 137-147.

Girardeau-Montaut, D., 2016. CloudCompare (Version 2.6. 0) [GPL Software].

Glen, R.A., Hancock, P.L., Whittaker, A., 2005. Basin inversion by distributed deformation: the southern margin of the Bristol Channel Basin, England. Journal of Structural Geology 27, 2113-2134.

Glennie, K.W., 1986. Development of N.W. Europe's southern Permian gas basin. Geological Society Special Publications 23, 3-22.

Glennie, K.W., 1997. Recent advances in understanding the southern North Sea Basin: a summary. Geological Society Special Publications 123, 17-29.

Grohmann, C.H., Campanha, G.A.C., 2010. OpenStereo: open source, cross-platform software for structural geology analysis, AGU Fall Meeting, San Francisco, CA, United States.

Gross, M.R., Fischer, M.P., Engelder, T., Greenfield, R.J., 1995. Factors controlling joint spacing in interbedded sedimentary rocks: integrating numerical models with field observations from the Monterey Formation, USA. Geological Society, London, Special Publications 92, 215-233.

Guerriero, V., Vitale, S., Ciarcia, S., Mazzoli, S., 2011. Improved statistical multi-scale analysis of fractured reservoir analogues. Tectonophysics 504, 14-24.

Haarmann, E., 1909. Die geologischen Verhältnisse des Piesbergsattels bei Osnabrück. Jahrbuch der Königl. Preuss. Geologischen Landesanstalt 30, 1-58.

Häfner, F., 2006. Simulation of the production behaviour of hydraulically fractured wells in tight gas reservoirs, DGMK-Report 593-9-2, Tight Gas Reservoirs. Deutsche Wissenschaftliche Gesellschaft für Erdöl, Erdgas und Kohle e.V., Hamburg, Germany, p. 236.

Hantschel, T., Kauerauf, A.I., 2009. Fundamentals of basin and petroleum systems modeling. Springer Verlag, Berlin, Germany.

Hardebol, N.J., Bertotti, G., 2013. DigiFract: A software and data model implementation for flexible acquisition and processing of fracture data from outcrops. Computers & Geosciences 54, 326-336.

Harris, N.B., Peters, K.E., 2012. Analyzing thermal histories of sedimentary basins: methods and case studies - introduction. Special Publication - Society for Sedimentary Geology 103, 1-4.

Harrison, T.M., Célérier, J., Aikman, A.B., Hermann, J., Heizler, M.T., 2009. Diffusion of 40Ar in muscovite. Geochimica et Cosmochimica Acta 73, 1039-1051.

Hartlieb, P., Toifl, M., Kuchar, F., Meisels, R., Antretter, T., 2016. Thermo-physical properties of selected hard rocks and their relation to microwave-assisted comminution. Minerals Engineering 91, 34-41.

Haszeldine, R.S., Samson, I.M., Cornford, C., 1984. Quartz diagenesis and convective fluid movement: Beatrice Oilfield, UK North Sea. Clay Minerals 19, 391-402.

Hedemann, H.A., Schuster, A., Stancu-Kristoff, G., Lösch, J., 1984. Die Verbreitung der Kohlenflöze des Oberkarbons in Nordwestdeutschland und ihre stratigraphische Einstufung. Fortschritte in der Geologie von Rheinland und Westfalen 32, 39-88.

Heidbach, O., Tingay, M., Barth, A., Reinecker, J., Kurfeß, D., Müller, B., 2008. The World Stress Map database release 2008.

Hinze, C., 1979. Erläuterungen zu Blatt Nr. 3614 Wallenhorst, Geologische Karte von Niedersachsen 1:25 000. Niedersächsisches Landesamt Bodenforschung, Hannover, Federal Republic of Germany.

Hitchmough, A.M., Riley, M.S., Herbert, A.W., Tellam, J.H., 2007. Estimating the hydraulic properties of the fracture network in a sandstone aquifer. Journal of contaminant hydrology 93, 38-57.

Hodgetts, D., 2013. Laser scanning and digital outcrop geology in the petroleum industry: A review. Marine and Petroleum Geology 46, 335-354.

Hoefs, J., 2009. Stable isotope geochemistry. Springer-Verlag, Berlin, Germany.

Hoffmann, N., Hengesbach, L., Friedrichs, B., Brink, H.J., 2008. The contribution of magnetotellurics to an improved understanding of the geological evolution of the North German Basin – review and new results. Zeitschrift der Deutschen Gesellschaft fuer Geowissenschaften 159, 591-606.

Hoffmann, N., Joedicke, H., Horejschi, L., 2005. Regional distribution of the Lower Carboniferous Culm and Carboniferous limestone facies in the North German Basin - derived from magnetotelluric soundings. Zeitschrift der Deutschen Gesellschaft fuer Geowissenschaften 156, 323-339.

Holditch, S.A., 2006. Tight gas sands. Journal of Petroleum Technology 58, 86.

Hollmann, F., Hülsmann, K.H., Schöne-Warnefeld, G., 1971. Die Karbonscholle des Piesberges, Internationaler Kongress für Stratigraphie und Geologie des Oberkarbons, Krefeld, Federal Republic of Germany.

Hollmann, G., Klug, B., Schmitz, J., Stahl, E., Wellens, M., 1998. Schneeren-Husum - zur Geologie einer Erdgaslagerstaette im nordwestdeutschen Oberkarbon. Forschung, Technik und Innovation - Preussag AG 23, 53-61.

Hotelling, H., 1933. Analysis of a complex of statistical variables into principal components. Journal of Educational Psychology 24, 417-441.

Houseknecht, D.W., Bensley, D.F., Hathon, L.A., Kastens, P.H., 1993. Rotational reflectance properties of Arkoma Basin dispersed vitrinite: insights for understanding reflectance populations in high thermal maturity regions. Organic Geochemistry 20, 187-196.

Houseknecht, D.W., Weesner, C.M.B., 1997. Rotational reflectance of dispersed vitrinite from the Arkoma basin. Organic Geochemistry 26, 191-206.

Howarth, R.J., 1998. Improved estimators of uncertainty in proportions, point-counting, and pass-fail test results. American Journal of Science 298, 594-607.

Hunziker, J.C., Frey, M., Clauer, N., Dallmeyer, R.D., Friedrichsen, H., Flehmig, W., Hochstrasser, K., Roggwiler, P., Schwander, H., 1986. The evolution of illite to muscovite: mineralogical and isotopic data from the Glarus Alps, Switzerland. Contributions to Mineralogy and Petrology 92, 157-180.

Jaboyedoff, M., Metzger, R., Oppikofer, T., Couture, R., Derron, M.H., Locat, J., Turmel, D., 2007. New insight techniques to analyze rock-slope relief using DEM and 3D-imaging cloud points, Rock Mechanics: Meeting Society's Challenges and Demands, Two Volume Set. Taylor & Francis, pp. 61-68.

Jacquemyn, C., Huysmans, M., Hunt, D., Casini, G., Swennen, R., 2015. Multi-scale three-dimensional distribution of fracture- and igneous intrusion-controlled hydrothermal dolomite from digital outcrop model, Latemar platform, Dolomites, northern Italy. AAPG Bulletin 99, 957-984.

Jäger, E., 1979. Introduction to geochronology, in: Jäger, E., Hunziker, J.C. (Eds.), Lectures in Isotope Geology. Springer-Verlag, Berlin, Germany, pp. 1-12.

Jahren, J.S., Aagaard, P., 1989. Compositional variations in diagenetic chlorites and illites, and relationships with formation-water chemistry. Clay Minerals 24, 157-170.

Johnson, S.C., 1967. Hierarchical clustering schemes. Psychometrika 32, 241-254.

Jones, N., Glover, B.W., 2005. Fluvial sandbody architecture, cyclicity and sequence stratigraphic setting: implications for hydrocarbon reservoirs - the Westphalian C and D of the Osnabruck-Ibbenburen area, northwest Germany, in: Collinson, J.D., Evans, D.J., Holliday, D.W., Jones, N.S. (Eds.), Carboniferous hydrocarbon geology: the southern North Sea and surrounding onshore areas. Yorkshire Geological Society, pp. 57-74.

Jones, R.R., McCaffrey, K.J.W., Clegg, P., Wilson, R.W., Holliman, N.S., Holdsworth, R.E., Imber, J., Waggott, S., 2009. Integration of regional to outcrop digital data: 3D visualisation of multi-scale geological models. Computers & Geosciences 35, 4-18.

Jones, R.R., Pringle, J.K., McCaffrey, K.J.W., Imber, J., Wightman, R.H., Guo, J., Long, J.J., 2011. Extending digital outcrop geology into the subsurface. Concepts in Sedimentology and Paleontology 10, 31-50.

Kantorowicz, J.D., 1985. The origin of authigenic ankerite from the Ninian Field, UK North Sea. Nature 315, 214-216.

Kemeny, J., Donovan, J., 2005. Rock mass characterization using lidar and automated point cloud processing. Ground Engineering 38, 26-29.

Kemeny, J., Turner, K., Norton, B., 2006. LIDAR for rock mass characterization: hardware, software, accuracy and best-practices. Laser and photogrammetric methods for rock face characterization, 49-62.

Kilby, W.E., 1988. Recognition of vitrinite with non-uniaxial negative reflectance characteristics. International Journal of Coal Geology 9, 267-285.

Klassen, H., 1984. Geologie des Osnabrücker Berglandes. Naturwissenschaftliches Museum Osnabrück, Germany.

Kley, J., Franzke, H.-J., Jähne, F., Krawczyk, C.M., Lohr, T., Reicherter, K., Scheck-Wenderoth, M., Sippel, J., Tanner, D., van Gent, H., 2008. Strain and stress, in: Littke, R., Bayer, U., Gajewski, D., Nelskamp, S. (Eds.), Dynamics of Complex Intracontinental Basins. Springer Verlag, Berlin, Germany, pp. 97-124.

Kley, J., Voigt, T., 2008. Late Cretaceous intraplate thrusting in Central Europe: effect of Africa-Iberia-Europe convergence, not Alpine collision. Geology 36, 839-842.

Kombrink, H., Besly, B.M., Collinson, J.D., Den Hartog Jager, D.G., Drozdzewski, G., Dusar, M., Hoth, P., Pagnier, H.J.M., Stemmerik, L., Waksmundzka, M.I., Wrede, V., 2010. Carboniferous, in: Doornebal, J.C., Stevenson, A.G. (Eds.), Petroleum Geological Atlas of the Southern Permian Basin Area. EAGE Publications b.v., Houten, pp. 81-99.

Koptíková, L., Bábek, O., Hladil, J., Kalvoda, J., Slavík, L., 2010. Stratigraphic significance and resolution of spectral reflectance logs in Lower Devonian carbonates of the Barrandian area, Czech Republic; a correlation with magnetic susceptibility and gamma-ray logs. Sedimentary Geology 225, 83-98.

Köstler, A.G., Ehrmann, W.U., 1991. Description of brittle extensional features in chalk on the crest of a salt ridge (NW Germany). Geological Society, London, Special Publications 56, 113-123.

Köwing, K., Rabitz, A., 2005. Osnabrücker Karbon, in: Amler, M.R.W., Wrede, V. (Eds.), Das Oberkarbon (Pennsylvanium) in Deutschland. Deutsche Stratigraphische Kommission, Stuttgart, Federal Republic of Germany, pp. 255-270.

Krooss, B.M., Plessen, B., Machel, H.G., Lüders, V., Littke, R., 2008. Origin and distribution of non-hydrocarbon gases, in: Littke, R., Bayer, U., Gajewski, D., Nelskamp, S. (Eds.), Dynamics of Complex Intracontinental Basins. Springer Verlag, Berlin, Germany, pp. 433-458.

Labourdette, R., Jones, R.R., 2007. Characterization of fluvial architectural elements using a three-dimensional outcrop data set: Escanilla braided system, South-Central Pyrenees, Spain. Geosphere 3, 422-434.

Lampe, C., Person, M.A., 2000. Episodic hydrothermal fluid flow in the upper Rhinegraben (Germany). Journal of Geochemical Exploration 69-70, 37-40.

Lampe, C., Schwark, L., 2012. Using geochemical analysis to identify a spatial maturity anomaly - an example from the Upper Rhine Graben, Germany. Special Publication - Society for Sedimentary Geology 103, 187-198.

Lanari, P., Wagner, T., Vidal, O., 2014. A thermodynamic model for di-trioctahedral chlorite from experimental and natural data in the system MgO–FeO–Al2O3–SiO2–H2O: applications to P–T sections and geothermometry. Contributions to Mineralogy and Petrology 167.

Larsen, B., Gudmundsson, A., Grunnaleite, I., Sælen, G., Talbot, M.R., Buckley, S.J., 2010. Effects of sedimentary interfaces on fracture pattern, linkage, and cluster formation in peritidal carbonate rocks. Marine and Petroleum Geology 27, 1531-1550.

Lato, M., Diederichs, M.S., Hutchinson, D.J., Harrap, R., 2009. Optimization of LiDAR scanning and processing for automated structural evaluation of discontinuities in rockmasses. International Journal of Rock Mechanics and Mining Sciences 46, 194-199.

Lato, M.J., Diederichs, M.S., Hutchinson, D.J., 2010. Bias Correction for View-limited Lidar Scanning of Rock Outcrops for Structural Characterization. Rock Mechanics and Rock Engineering 43, 615-628.

Lato, M.J., Vöge, M., 2012. Automated mapping of rock discontinuities in 3D lidar and photogrammetry models. International Journal of Rock Mechanics and Mining Sciences 54, 150-158.

Laubach, S.E., Olson, J.E., Eichhubl, P., 2009. Fracture diagenesis and producibility in tight gas sandstones, in: Carr, T., D'Agostino, T., Ambrose, W., Pashin, J., Rosen, N.C. (Eds.), Unconventional Energy Resources: Making the Unconventional Conventional, 29th Annual Bob F. Perkins Research Conference. Society of Economic Paleontologists, Gulf Coast Section (GCSSEPM) Foundation, Houston, TX, United States, pp. 483-499.

Laux, D., Henk, A., 2015. Terrestrial laser scanning and fracture network characterisation – perspectives for a (semi-) automatic analysis of point cloud data from outcrops. Zeitschrift der deutschen Gesellschaft für Geowissenschaften 166, 99-118.

Law, B.E., 2002. Basin-centered gas systems. AAPG Bulletin 86, 1891-1919

Law, B.E., Curtis, J.B., 2002. Unconventional petroleum systems. American Association of Petroleum Geologists : Tulsa, OK, United States, United States, pp. 1851-2005.

Law, B.E., Spencer, C.W., Charpentier, R.R., Crovelli, R.A., Mast, R.F., Dolton, G.L., Wandrey, C.J., 1989. Estimates of gas resources in overpressured low-permeability Cretaceous and Tertiary sandstone reservoirs, greater Green River basin, Wyoming, Colorado, and Utah. Guidebook - Wyoming Geological Association 40, 39-61.

Law, B.E., Ulmishek, G.F., Clayton, J.L., Kabyshev, B.P., Pashova, N.T., Krivoshey, V.A., 1998. Basin-centered gas evaluated in Dnieper-Donets Basin, Donbas Foldbelt, Ukraine. Oil & Gas Journal 96, 74-78.

Le Garzic, E., de L'Hamaide, T., Diraison, M., Géraud, Y., Sausse, J., de Urreiztieta, M., Hauville, B., Champanhet, J.-M., 2011. Scaling and geometric properties of extensional fracture systems in the proterozoic basement of Yemen. Tectonic interpretation and fluid flow implications. Journal of Structural Geology 33, 519-536.

Leeder, M.R., Hardman, M., 1990. Carboniferous geology of the Southern North Sea Basin and controls on hydrocarbon prospectivity. Geological Society, London, Special Publications 55, 87-105.

Lemoalle, J., Dupont, B., 1973. Iron-bearing Oolites and the Present Conditions of Iron Sedimentation in Lake Chad (Africa). International Union of Geological Sciences. Series A 3, 167-178.

Levine, J.R., Davis, A., 1984. Optical anisotropy of coals as an indicator of tectonic deformation, Broad Top coal field, Pennsylvania. Geological Society of America Bulletin 95, 100-108.

Levine, J.R., Davis, A., 1989. Reflectance anisotropy of Upper Carboniferous coals in the Appalachian foreland basin, Pennsylvania, U.S.A. International Journal of Coal Geology 13, 341-373.

Li, Z., Chi, G., Bethune, K.M., 2016. The effects of basement faults on thermal convection and implications for the formation of unconformity-related uranium deposits in the Athabasca Basin, Canada. Geofluids 16, 729-751.

Liewig, N., Clauer, N., 2000. K-Ar dating of varied microtextural illite in Permian gas reservoirs, northern Germany. Clay Minerals 35, 271-281.

Littke, R., Bayer, U., Gajewski, D., Nelskamp, S., 2008. Dynamics of complex intracontinental basins. Springer, Berlin, Federal Republic of Germany.

Littke, R., Bueker, C., Hertle, M., Karg, H., Stroetmann-Heinen, V., Oncken, O., 2000. Heat flow evolution, subsidence and erosion in the Rheno-Hercynian orogenic wedge of Central Europe. Geological Society Special Publications 179, 231-255.

Littke, R., Krooss, B., Uffmann, A.K., Schulz, H.M., Horsfield, B., 2011. Unconventional gas resources in the Paleozoic of Central Europe. Oil & Gas Science and Technology 66, 953-977.

Littke, R., Urai, J.L., Uffmann, A.K., Risvanis, F., 2012. Reflectance of dispersed vitrinite in Palaeozoic rocks with and without cleavage: Implications for burial and thermal history modeling in the Devonian of Rursee area, northern Rhenish Massif, Germany. International Journal of Coal Geology 89, 41-50.

Lohr, T., Krawczyk, M., Tanner, D.C., Samiee, R., Endres, H., Oncken, O., Trappe, H., Kukla, P.A., 2007. Strain partitioning due to salt: insights from interpretation of a 3D seismic data set in the NW German Basin. Basin Research 19, 579-597.

Lotze, F., 1953. Das Alter der Erzvorkommen des Osnabrücker Gebiets im Verhältnis zur Tektonik. Neues Jahrbuch für Geologie und Paläontologie. Monatsheft 8, 336-342.

Lüders, V., Plessen, B., 2015. Stable carbon isotope ratios of CH4-rich gas inclusions in shale-hosted fracture-fill mineralization: A tool for tracing hydrocarbon generation and migration in shale plays for oil and gas. Marine and Petroleum Geology 63, 68-81.

Lüders, V., Plessen, B., di Primio, R., 2012. Stable carbon isotopic ratios of CH4–CO2-bearing fluid inclusions in fracture-fill mineralization from the Lower Saxony Basin (Germany) – A tool for tracing gas sources and maturity. Marine and Petroleum Geology 30, 174-183.

Lundegard, P.D., 1992. Sandstone porosity loss - a "big picture" view of the importance of compaction. Journal of Sedimentary Research 62, 250-260.

Magri, F., Littke, R., Rodon, S., Bayer, U., Urai, J.L., 2008. Temperature fields, petroleum maturation and fluid flow in the vicinity of salt domes, in: Littke, R., Bayer, U., Gajewski, D., Nelskamp, S. (Eds.), Dynamics of Complex Intracontinental Basins. Springer Verlag, Berlin, Germany, pp. 323-344.

Mah, J., Samson, C., McKinnon, S.D., 2011. 3D laser imaging for joint orientation analysis. International Journal of Rock Mechanics and Mining Sciences 48, 932-941.

Mahmud, K., Mariethoz, G., Treble, P.C., Baker, A., 2015. Terrestrial LiDAR Survey and Morphological Analysis to Identify Infiltration Properties in the Tamala Limestone, Western Australia. IEEE Journal of Selected Topics in Applied Earth Observations and Remote Sensing 8, 4871-4881.

Marrett, R., Ortega, O.J., Kelsey, C.M., 1999. Extent of power-law scaling for natural fractures in rock. Geology 27, 799-802.

Mauldon, M., Dunne, W.M., Rohrbaugh, M.B., 2001. Circular scanlines and circular windows: new tools for characterizing the geometry of fracture traces. Journal of Structural Geology 23, 247-258.

McBride, B.C., Weimer, P., Rowan, M.G., 1998. The effect of allochthonous salt on the petroleum systems of northern Green Canyon and Ewing Bank (offshore Louisiana), northern Gulf of Mexico. AAPG Bulletin 82, 1083-1112.

McBride, E.F., 1963. A classification of common sandstones. Journal of Sedimentary Petrology 33, 664-669.

McCaffrey, K.J.W., Jones, R.R., Holdsworth, R.E., Wilson, R.W., Clegg, P., Imber, J., Holliman, N., Trinks, I., 2005. Unlocking the spatial dimension: digital technologies and the future of geoscience fieldwork. Journal of the Geological Society 162, 927-938.

McKinley, J.M., Atkinson, P.M., Lloyd, C.D., Ruffell, A.H., Worden, R.H., 2011. How porosity and permeability vary spatially with grain size, sorting, cement volume, and mineral dissolution in fluvial Triassic sandstones: The value of geostatistics and local regression. Journal of Sedimentary Research 81, 844-858.

Meckel, L.D., Jr., Smith, D.G., Wells, L.A., 1992. Ouachita foredeep basins: Regional paleogeography and habitat of hydrocarbons. AAPG Memoir 55, 427-444.

Merriman, R.J., Peacor, D.R., 1999. Very low-grade metapelites. Mineralogy, microfabrics and measuring reaction progress, in: Frey, M., Robinson, D. (Eds.), Low-grade metamorphism. Blackwell Science, Oxford, UK, pp. 10-58.

Miall, A.D., 1988a. Architectural elements and bounding surfaces in fluvial deposits: anatomy of the Kayenta Formation (Lower Jurassic), southwest Colorado. Sedimentary Geology 55, 233-262.

Miall, A.D., 1988b. Facies Architecture in Clastic Sedimentary Basins, in: Kleinspehn, K.L., Paola, C. (Eds.), New Perspectives in Basin Analysis. Springer, New York, NY, United States, pp. 67-81.

Möbius, J., Kobbelt, L., 2012. OpenFlipper: An Open Source Geometry Processing and Rendering Framework, in: Boissonnat, J.-D., Chenin, P., Cohen, A., Gout, C., Lyche, T., Mazure, M.-L., Schumaker, L. (Eds.), Curves and Surfaces: 7th International Conference, Avignon, France, June 24 - 30, 2010, Revised Selected Papers. Springer Berlin Heidelberg, Berlin, Heidelberg, pp. 488-500.

Monsen, E., Hunt, D.W., Bounaim, A., Savary-Sismondini, B., Brenna, T., Nickel, M., Sonneland, L., Thurmond, J.B., Gillespie, P., 2011. The automated interpretation of photorealistic outcrop models. Concepts in Sedimentology and Paleontology 10, 107-135.

Morad, S., Al-Ramadan, K., Ketzer, J.M., De Ros, L.F., 2010. The impact of diagenesis on the heterogeneity of sandstone reservoirs: A review of the role of depositional facies and sequence stratigraphy. AAPG Bulletin 94, 1267-1309.

Morad, S., Ketzer, J.M., De Ros, L.F., 2000. Spatial and temporal distribution of diagenetic alterations in siliciclastic rocks: implications for mass transfer in sedimentary basins. Sedimentology 47, Suppl. 1, 95-120.

Moussa, M.T., 1973. Measuring volumes of sedimentary grains. Journal of Sedimentary Petrology 43, 1171-1173.

Nicol, A., Watterson, J., Walsh, J.J., Childs, C., 1996. The shapes, major axis orientations and displacement patterns of fault surfaces. Journal of Structural Geology 18, 235-248.

O'Brien, J.J., Lerche, I., 1988. Impact of heat flux anomalies around salt diapirs and salt sheets in the Gulf Coast on hydrocarbon maturity: models and observations. Transactions - Gulf Coast Association of Geological Societies 38, 231-243.

Odin, G.-S., Bonhomme, M.G., 1982. Argon behaviour in clays and glauconies during preheating experiments, in: Odin, G.-S. (Ed.), Numerical Dating in Stratigraphy. John Wiley & Sons, Chichester, United Kingdom, pp. 333-343.

Odling, N.E., Gillespie, P., Bourgine, B., Castaing, C., Chiles, J.P., Christensen, N.P., Fillion, E., Genter, A., Olsen, C., Thrane, L., Trice, R., Aarseth, E., Walsh, J.J., Watterson, J., 1999. Variations in fracture system geometry and their implications for fluid flow in fractured hydrocarbon reservoirs. Petroleum Geoscience 5, 373-384.

Odling, N.E., Harris, S.D., Knipe, R.J., 2004. Permeability scaling properties of fault damage zones in siliclastic rocks. Journal of Structural Geology 26, 1727-1747.

Okrusch, M., Matthes, S., 2014. Mineralogie: eine Einführung in die spezielle Mineralogie, Petrologie und Lagerstaettenkunde. Springer Spektrum, Berlin.

Olariu, C., Bhattacharya, J.P., Xu, X., Aiken, C.L.V., Zeng, X., McMechan, G.A., 2005. Integrated study of ancient delta-front deposits, using outcrop, ground-penetrating radar, and three-dimensional photorealistic data: Cretaceous Panther Tongue sandstone, Utah, U. S. A. Special Publication - Society for Sedimentary Geology 83, 155-177.

Olariu, M.I., Ferguson, J.F., Aiken, C.L.V., Xu, X., 2008. Outcrop fracture characterization using terrestrial laser scanners: Deep-water Jackfork sandstone at Big Rock Quarry, Arkansas. Geosphere 4, 247-259.

Olivarius, M., Weibel, R., Hjuler, M.L., Kristensen, L., Mathiesen, A., Nielsen, L.H., Kjoller, C., 2015. Diagenetic effects on porosity-permeability relationships in red beds of the Lower Triassic Bunter Sandstone Formation in the North German Basin. Sedimentary Geology 321, 139-153.

Olson, J.E., Laubach, S.E., Lander, R.H., 2009. Natural fracture characterization in tight gas sandstones: Integrating mechanics and diagenesis. AAPG Bulletin 93, 1535-1549.

Ortega, O., Marrett, R., 2000. Prediction of macrofracture properties using microfracture information, Mesaverde Group sandstones, San Juan basin, New Mexico. Journal of Structural Geology 22, 571-588.

Ortega, O.J., Marrett, R.A., Laubach, S.E., 2006. A scale-independent approach to fracture intensity and average spacing measurement. AAPG Bulletin 90, 193-208.

Paxton, S.T., Szabo, J.O., Ajdukiewicz, J.M., 2002. Construction of an intergranular volume compaction curve for evaluating and predicting compaction and porosity loss in rigid-grain sandstone reservoirs. AAPG Bulletin 86, 2045-2065.

Petmecky, S., Meier, L., Reiser, H., Littke, R., 1999. High thermal maturity in the Lower Saxony Basin: intrusion or deep burial? Tectonophysics 304, 317-344.

Pfeifer, N., Höfle, B., Briese, C., Rutzinger, M., Haring, A., 2008. Analysis of the backscattered energy in terrestrial laser scanning data. Int. Arch. Photogramm. Remote Sens. Spat. Inf. Sci 37, 1045-1052.

Pless, J.C., McCaffrey, K.J.W., Jones, R.R., Holdsworth, R.E., Conway, A., Krabbendam, M., 2015. 3D characterization of fracture systems using Terrestrial Laser Scanning. Geological Society, London, Special Publications 421, 125-141.

Plessen, B., Lüders, V., 2012. Simultaneous measurements of gas isotopic compositions of fluid inclusion gases (N2, CH4, CO2) using continuous-flow isotope ratio mass spectrometry. Rapid Communications in Mass Spectrometry 26, 1157-1161.

Pollard, D.D., Aydin, A., 1988. Progress in understanding jointing over the past century. GSA Bulletin 100, 1181-1204.

Popov, M.A., Nuccio, V.F., Dyman, T.S., Gognat, T.A., Johnson, R.C., Schmoker, J.W., Wilson, M.S., Bartherger, C., 2001. Basin-centered gas systems of the U.S. Geological Survey, Reston, VA, United States.

Porrenga, D.H., 1965. Chamosite in recent sediments of the Niger and Orinoco deltas. Geologie en Mijnbouw 44, 400-403.

Pranter, M.J., Ellison, A.I., Cole, R.D., Patterson, P.E., 2007. Analysis and modeling of intermediate-scale reservoir heterogeneity based on a fluvial point-bar outcrop analog, Williams Fork Formation, Piceance Basin, Colorado. AAPG Bulletin 91, 1025-1051.

Pranter, M.J., Hewlett, A.C., Cole, R.D., Wang, H., Gilman, J., 2014. Fluvial architecture and connectivity of the Williams Fork Formation: use of outcrop analogues for stratigraphic characterization and reservoir modelling. Geological Society, London, Special Publications 387, 57-83.

Priest, S.D., Hudson, J.A., 1981. Estimation of discontinuity spacing and trace length using scanline surveys. International Journal of Rock Mechanics and Mining Sciences & Geomechanics Abstracts 18, 183-197.

Pringle, J.K., Brunt, R.L., Hodgson, D.M., Flint, S.S., 2010. Capturing stratigraphic and sedimentological complexity from submarine channel complex outcrops to digital 3D models, Karoo Basin, South Africa. Petroleum Geoscience 16, 307-330.

Pringle, J.K., Howell, J.A., Hodgetts, D., Westerman, A.R., Hodgson, D.M., 2006. Virtual outcrop models of petroleum reservoir analogues: a review of the current state-of-the-art. First Break 24, 33-42.

Purdy, J.W., Jäger, E., 1976. K-Ar Ages on Rock Forming Minerals from the Central Alps. University of Padova, Italy.

Pusch, G., Gaupp, R., Liermann, N., 2005. Integrated research contributions for screening the Tight Gas Potential in the Rotliegendes Formations of North-Germany. Oil Gas European Magazine 31, 187-193.

Rarity, F., van Lanen, X.M.T., Hodgetts, D., Gawthorpe, R.L., Wilson, P., Fabuel-Perez, I., Redfern, J., 2014. LiDAR-based digital outcrops for sedimentological analysis: workflows and techniques. Geological Society, London, Special Publications 387, 153-183.

Rawnsley, K.D., Rives, T., Petti, J.P., Hencher, S.R., Lumsden, A.C., 1992. Joint development in perturbed stress fields near faults. Journal of Structural Geology 14, 939-951.

Redfern, J., Hodgetts, D., Fabuel-Perez, I., 2007. Digital analysis brings renaissance for petroleum geology outcrop studies in North Africa. First Break 25, 81-87.

Reich, H., 1933. Erdmagnetismus und saxonische Tektonik. Zeitschrift der Deutschen Geologischen Gesellschaft 85, 635-646.

Reutel, C., Lüders, V., Hoth, P., Idiz, E.F., 1995. Gas migration and accumulation along lineament structures - Lower Saxony Basin (NW Germany). Boletín de la Sociedad Española de Mineralogía 18, 205-206.

Rezaee, R., Saeedi, A., Clennell, B., 2012. Tight gas sands permeability estimation from mercury injection capillary pressure and nuclear magnetic resonance data. Journal of Petroleum Science and Engineering 88-89, 92-99.

Rider, M.H., Kennedy, M., 2011. The Geological Interpretation of Well Logs, 3rd ed. Rider-French Consulting Ltd.

Riquelme, A.J., Abellán, A., Tomás, R., Jaboyedoff, M., 2014. A new approach for semi-automatic rock mass joints recognition from 3D point clouds. Computers & Geosciences 68, 38-52.

Robinson, A.G., Coleman, M.L., Gluyas, J.G., 1993. The age of illite cement growth, Village Fields area, southern North Sea: Evidence from K-Ar ages and 18O/16O ratios. AAPG Bulletin 77, 68-80.

Roedder, E., 1984. Fluid inclusions. Mineralogical Society of America, Washington, DC, United States.

Rogers, J.J.W., Head, W.B., 1961. Relationships between porosity, median size, and sorting coefficients of synthetic sands. Journal of Sedimentary Petrology 31, 467-470.

Röhrs, H., 1992. Erz und Kohle - Bergbau und Eisenhütten zwischen Ems und Weser. Ibbenbürener Vereinsdruckerei, Ibbenbüren, Germany.

Roncella, R., Forlani, G., 2005. Extraction of planar patches from point clouds to retrieve dip and dip direction of rock discontinuities. The International Archives of the Photogrammetry, Remote Sensing and Spatial Information Sciences 36, 162-167.

Roscher, M., Schneider, J.W., 2006. Permo-Carboniferous climate: Early Pennsylvanian to Late Permian climate development of central Europe in a regional and global context. Geological Society, London, Special Publications 265, 95-136.

Rotevatn, A., Buckley, S.J., Howell, J.A., Fossen, H., 2009. Overlapping faults and their effect on fluid flow in different reservoir types: A LIDAR-based outcrop modeling and flow simulation study. AAPG Bulletin 93, 407-427.

Sakhaee-Pour, A., Bryant, S.L., 2014. Effect of pore structure on the producibility of tight-gas sandstones. AAPG Bulletin 98, 663-694.

Santos, R.F.V.C., Miranda, T.S., Barbosa, J.A., Gomes, I.F., Matos, G.C., Gale, J.F.W., Neumann, V.H.L.M., Guimaraes, L.J.N., 2015. Characterization of natural fracture systems: Analysis of uncertainty effects in linear scanline results. AAPG Bulletin 99, 2203-2219.

Scheck, M., Bayer, U., 1999. Evolution of the Northeast German Basin - inferences from a 3D structural model and subsidence analysis. Tectonophysics 313, 145-169.

Schegg, R., Leu, W., Greber, E., 1997. New exploration concepts spark Swiss gas, oil prospects. Oil & Gas Journal 95, 102-106.

Scheidt, G., Littke, R., 1989. Comparative organic petrology of interlayered sandstones, siltstones, mudstones and coals in the Upper Carboniferous Ruhr Basin, Northwest Germany, and their thermal history and methane generation. Geologische Rundschau 78, 375-390.

Schlische, R.W., Young, S.S., Ackermann, R.V., Gupta, A., 1996. Geometry and scaling relations of a population of very small rift-related normal faults. Geology 24, 683-686.

Schmidt, V., McDonald, D.A., 1979. The role of secondary porosity in the course of sandstone diagenesis. SEPM Special Publication 26, 175-207.

Schöner, R., Gaupp, R., 2008. Diagenesis of Rotliegend red bed sandstones in the North German Basin: Implications for fluid evolution and migration. Schriftenreihe der Deutschen Gesellschaft für Geowissenschaften 60, 111-111.

Schultz, R.A., Fossen, H., 2002. Displacement-length scaling in three dimensions: the importance of aspect ratio and application to deformation bands. Journal of Structural Geology 24, 1389-1411.

Schultz, R.A., Okubo, C.H., Wilkins, S.J., 2006. Displacement-length scaling relations for faults on the terrestrial planets. Journal of Structural Geology 28, 2182-2193.

Sedat, B., 1992. Petrographie und Diagenese von Sandsteinen im Nordwestdeutschen Oberkarbon, DGMK-Bericht 384-7, Sedimentologie des Oberkarbons. Deutsche Gesellschaft für Mineralölwissenschaft und Kohlechemie e.V., Hamburg, Germany, p. 143.

Seers, T.D., Hodgetts, D., 2013. Comparison of digital outcrop and conventional data collection approaches for the characterization of naturally fractured reservoir analogues. Geological Society, London, Special Publications 374, 51-77.

Seers, T.D., Hodgetts, D., 2016a. Extraction of three-dimensional fracture trace maps from calibrated image sequences. Geosphere 12, 1323-1340.

Seers, T.D., Hodgetts, D., 2016b. Probabilistic constraints on structural lineament best fit plane precision obtained through numerical analysis. Journal of Structural Geology 82, 37-47.

Senglaub, Y., Brix, M.R., Adriasola, A.C., Littke, R., 2005. New information on the thermal history of the southwestern Lower Saxony Basin, northern Germany, based on fission track analysis. International Journal of Earth Sciences 94, 876-896.

Senglaub, Y., Littke, R., Brix, M.R., 2006. Numerical modelling of burial and temperature history as an approach for an alternative interpretation of the Bramsche anomaly, Lower Saxony Basin. International Journal of Earth Sciences 95, 204-224.

Sharp, J.M., Jr., Shi, M., Galloway, W.E., 2003. Heterogeneity of fluvial systems - control on density-driven flow and transport. Environmental & Engineering Geoscience 9, 5-17.

Shipton, Z.K., Soden, A.M., Kirkpatrick, J.D., Bright, A.M., Lunn, R.J., 2006. How Thick is a Fault? Fault Displacement-Thickness Scaling Revisited. Earthquakes: Radiated Energy and the Physics of Faulting, 193-198.

Simms, M.A., Garven, G., 2004. Thermal convection in faulted extensional sedimentary basins: theoretical results from finite-element modeling. Geofluids 4, 109-130.

Sippel, J., Scheck-Wenderoth, M., Reicherter, K., Mazur, S., 2009. Paleostress states at the south-western margin of the Central European Basin System — Application of fault-slip analysis to unravel a polyphase deformation pattern. Tectonophysics 470, 129-146.

Soliva, R., Schultz, R.A., Benedicto, A., 2005. Three-dimensional displacement-length scaling and maximum dimension of normal faults in layered rocks. Geophysical Research Letters 32.

Song, J.-H., Han, S.-H., Yu, K., Kim, Y.-I., 2002. Assessing the possibility of land-cover classification using lidar intensity data. International Archives of Photogrammetry Remote Sensing and Spatial Information Sciences 34, 259-262.

Spencer, C.W., Mast, R.F., 1986. Geology of tight gas reservoirs. American Association of Petroleum Geologists, Tulsa, OK, United States.

Spoetl, C., Houseknecht, D.W., Longstaffe, F.J., 1994. Authigenic chlorites in sandstones as indicators of high-temperature diagenesis, Arkoma foreland basin, USA. Journal of Sedimentary Research (A) 64, 553-566.

Stadler, G., Teichmüller, R., 1971. Zusammenfassender Überblick über die Entwicklung des Bramscher Massivs und des Niedersächsischen Tektogens. Fortschritte in der Geologie von Rheinland und Westfalen 18, 547-564.

Stahl, W., 1971. Isotopen-Analysen an Carbonaten und Kohlendioxid-Proben aus dem Einflussbereich und der weiteren Umgebung des Bramscher Intrusivs und an hydrothermalen Carbonaten aus dem Siegerland. Fortschritte in der Geologie von Rheinland und Westfalen 18, 430-438.

Stahl, W.J., 1977. Carbon and nitrogen isotopes in hydrocarbon research and exploration. Chemical Geology 20, 121-149.

Steiger, R.H., Jäger, E., 1977. Subcommission on geochronology: Convention on the use of decay constants in geo- and cosmochronology. Earth and Planetary Science Letters 36, 359-362.

Strijker, G., Bertotti, G., Luthi, S.M., 2012. Multi-scale fracture network analysis from an outcrop analogue: A case study from the Cambro-Ordovician clastic succession in Petra, Jordan. Marine and Petroleum Geology 38, 104-116.

Stroker, T.M., Harris, N.B., Crawford Elliott, W., Marion Wampler, J., 2013. Diagenesis of a tight gas sand reservoir: Upper Cretaceous Mesaverde Group, Piceance Basin, Colorado. Marine and Petroleum Geology 40, 48-68.

Sturzenegger, M., Stead, D., 2009. Quantifying discontinuity orientation and persistence on high mountain rock slopes and large landslides using terrestrial remote sensing techniques. Natural Hazards and Earth System Sciences 9, 267-287.

Süss, M.P., Drozdzewski, G., Schäfer, A., 2000. Sequenzstratigraphie des kohleführenden Oberkarbons im Ruhr-Becken. Geologisches Jahrbuch A 156, 45-106.

Tavani, S., Corradetti, A., Billi, A., 2016. High precision analysis of an embryonic extensional fault-related fold using 3D orthorectified virtual outcrops: The viewpoint importance in structural geology. Journal of Structural Geology 86, 200-210.

Teichmüller, M., 1986. Coalification and natural gas deposits in northwestern Germany. Geological Society, London, Special Publications 23, 101-112.

Teichmüller, M., Teichmüller, R., Bartenstein, H., 1984. Inkohlung und Erdgas - eine neue Inkohlungskarte der Karbon-Oberfläche in Nordwestdeutschland. Fortschritte in der Geologie von Rheinland und Westfalen 32, 11-34.

Terzaghi, R.D., 1965. Sources of Error in Joint Surveys. Géotechnique 15, 287-304.

Thiery, R., Vidal, J., Dubessy, J., 1994. Phase equilibria modelling applied to fluid inclusions: liquid–vapour equilibria and calculation of the molar volume in the CO2–CH4–N2 system. Geochimica et Cosmochimica Acta 58, 1073-1082.

Timar-Geng, Z., Fügenschuh, B., Schaltegger, U., Wetzel, A., 2004. The impact of the Jurassic hydrothermal activity on zircon fission track data from the southern Upper Rhine Graben area. Schweizerische Mineralogische und Petrographische Mitteilungen 84, 257-269.

Timar-Geng, Z., Henk, A., Wetzel, A., 2009. Convective heat transfer in a steeply dipping fault zone and its impact on the interpretation of fission-track data - a modelling study. Geological Society Special Publications 324, 87-98.

Torabi, A., Berg, S.S., 2011. Scaling of fault attributes: A review. Marine and Petroleum Geology 28, 1444-1460.

Tucker, M.E., 1991. Sedimentary petrology: an introduction to the origin of sedimentary rocks. Blackwell Scientific Publishing, Oxford, United Kingdom.

Turcotte, D.L., Schubert, G., 2002. Geodynamics. Cambridge University Press, Cambridge, United Kingdom.

Turner, P., Burley, S.D., Rey, D., Prosser, J., 1995. Burial history of the Penrith Sandstone (Lower Permian) deduced from the combined study of fluid inclusion and palaeomagnetic data. Geological Society, London, Special Publications 98, 43-78.

Ufer, K., Stanjek, H., Roth, G., Dohrmann, R., Kleeberg, R., Kaufhold, S., 2008. Quantitative phase analysis of bentonites by the Rietveld method. Clays and Clay Minerals 56, 272-282.

Umili, G., Ferrero, A., Einstein, H.H., 2013. A new method for automatic discontinuity traces sampling on rock mass 3D model. Computers & Geosciences 51, 182-192.

van Knapen, B., Slob, S., 2006. Identification and characterisation of rock mass discontinuity sets using 3D laser scanning. Engineering Geology for Tomorrow's Cities: Geological Society, London, Engineering Geology Special Publications 22.

van Wees, J.D., Stephenson, R.A., Ziegler, P.A., Bayer, U., McCann, T., Dadlez, R., Gaupp, R., Narkiewicz, M., Bitzer, F., Scheck, M., 2000. On the origin of the Southern Permian Basin, Central Europe. Marine and Petroleum Geology 17, 43-59.

Vasuki, Y., Holden, E.-J., Kovesi, P., Micklethwaite, S., 2014. Semi-automatic mapping of geological Structures using UAV-based photogrammetric data: An image analysis approach. Computers & Geosciences 69, 22-32.

Vöge, M., Lato, M.J., Diederichs, M.S., 2013. Automated rockmass discontinuity mapping from 3-dimensional surface data. Engineering Geology 164, 155-162.

Voigt, E., 1960. Über den Zechsteingraben des Piesberges bei Osnabrück. Mitteilungen aus dem Geologischen Staatsinstitut in Hamburg 29, 5-25.

von Flotow, A., Berroth, A., Schmehl, H., Kossmat, F., 1931. Relative Bestimmung der Schwerkraft auf 115 Stationen in Norddeutschland. Veröffentlichung des Preussischen Geodätischen Instituts, N.F. 106, 1-88.

Voyat, I., Roncella, R., Forlani, G., Ferrero, A.M., 2006. Advanced techniques for geo structural surveys in modelling fractured rock masses: application to two Alpine sites. American Rock Mechanics Association.

Wang, X., 2005, Stereological interpretation of rock fracture traces on borehole walls and other cylindrical surfaces. PhD thesis, Virginia State University, United States.

Welte, D.H., Yalçin, M.N., 1988. Basin modelling—A new comprehensive method in petroleum geology. Organic Geochemistry 13, 141-151.

Wentworth, C.K., 1922. A scale of grade and class terms for clastic sediments. Journal of Geology 30, 377-392.

Wilkinson, M., Haszeldine, R.S., 2002. Fibrous illite in oilfield sandstones - a nucleation kinetic theory of growth. Terra Nova 14, 56-60.

Will, P., Lüders, V., Wemmer, K., Gilg, H.A., 2016. Pyrophyllite formation in the thermal aureole of a hydrothermal system in the Lower Saxony Basin, Germany. Geofluids 16, 349-363.

Wilson, C.E., Aydin, A., Karimi-Fard, M., Durlofsky, L.J., Sagy, A., Brodsky, E.E., Kreylos, O., Kellogg, L.H., 2011. From outcrop to flow simulation: Constructing discrete fracture models from a LIDAR survey. AAPG Bulletin 95, 1883-1905.

Wilson, P., Hodgetts, D., Rarity, F., Gawthorpe, R.L., Sharp, I.R., 2009. Structural geology and 4D evolution of a half-graben: New digital outcrop modelling techniques applied to the Nukhul half-graben, Suez rift, Egypt. Journal of Structural Geology 31, 328-345.

Wilson, R.W., McCaffrey, K.J.W., Holdsworth, R.E., Imber, J., Jones, R.R., Welbon, A.I.F., Roberts, D., 2006. Complex fault patterns, transtension and structural segmentation of the Lofoten Ridge, Norwegian margin: Using digital mapping to link onshore and offshore geology. Tectonics 25.

Wimmers, K., Koehrer, B., 2014. Integration of Sedimentology, Petrophysics and Rock Typing as Key to Understanding a Tight Gas Reservoir. OIL GAS European Magazine 40.

Wisian, K.W., Blackwell, D.D., 2004. Numerical modeling of Basin and Range geothermal systems. Geothermics 33, 713-741.

Woodcock, N.H., 1977. Specification of fabric shapes using an eigenvalue method. GSA Bulletin 88, 1231-1236.

Worden, R.H., Morad, S., 2003. Clay minerals in sandstones: controls on formation, distribution and evolution. Special Publication of the International Association of Sedimentologists 34, 3-41.

Wüstefeld, P., Hilgers, C., Koehrer, B., Höhne, M., Steindorf, P., Schurk, K., Becker, S., Bertier, P., 2014. Reservoir Heterogeneity in Upper Carboniferous Tight Gas Sandstones: Lessons Learned From an Analog Study, SPE/EAGE European Unconventional Resources Conference and Exhibition 2014. Society of Petroleum Engineers, Vienna, Austria.

Wüstefeld, P., Hilse, U., Koehrer, B., Adelmann, D., Hilgers, C., 2017a. Critical evaluation of an Upper Carboniferous tight gas sandstone reservoir analog: Diagenesis and petrophysical aspects. Marine and Petroleum Geology 86, 689-710.

Wüstefeld, P., Hilse, U., Lüders, V., Wemmer, K., Koehrer, B., Hilgers, C., 2017b. Kilometer-scale fault-related thermal anomalies in tight gas sandstones. Marine and Petroleum Geology 86, 288-303.

Xu, X., Bhattacharya, J.B., Davies, R.K., Aiken, C.L.V., 2001. Digital Geologic Mapping of the Ferron Sandstone, Muddy Creek, Utah, with GPS and Reflectorless Laser Rangefinders. GPS Solutions 5, 15-23.

Yalçin, M.N., Littke, R., Sachsenhofer, R.F., 1997. Thermal history of sedimentary basins. Springer Verlag, Berlin, Germany, pp. 71-167.

Yale, D.P., 2003. Fault and stress magnitude controls on variations in the orientation of in situ stress. Geological Society, London, Special Publications 209, 55-64.

Ziegler, P.A., 1977. Geology and hydrocarbon provinces of the North Sea. GeoJournal 1, 7-32.

Ziegler, P.A., 1982. Triassic rifts and facies patterns in Western and Central Europe. Geologische Rundschau 71, 747-772.

Ziegler, P.A., 1987. Late Cretaceous and Cenozoic intra-plate compressional deformations in the Alpine foreland—a geodynamic model. Tectonophysics 137, 389-420.

Zwingmann, H., Clauer, N., Gaupp, R., 1998. Timing of fluid flow in a sandstone reservoir of the North German Rotliegend (Permian) by K-Ar dating of related hydrothermal illite. Geological Society Special Publications 144, 91-106.

Zwingmann, H., Clauer, N., Gaupp, R., 1999. Structure-related geochemical (REE) and isotopic (K-Ar, Rb-Sr, δ18O) characteristics of clay minerals from Rotliegend sandstone reservoirs (Permian, northern Germany). Geochimica et Cosmochimica Acta 63, 2805-2823.

Appendices

A1 Ground truthing

Ground truthing measurements were carried out in the Piesberg quarry and the nearby Woitzel outcrop. The average deviation of the digitally acquired orientation data is 2.1° for both the dip direction and the dip (Tab A. 1). Windsor & Robinson (1994) describe average uncertainties of compass measurements of ± 2.5° vertical and ± 7.5 azimuthal error. The outlined approach for the digital fracture detection is assumed as valid option for the derivation of fracture statistics because the average deviations are within this measuring inaccuracy of compass data.

Table A.1: Ground truthing measurements, comparing manual data derived with a geological compass with digitally derived orientation measures by the outlined method in chapter 4. The underscored deviation values between both methods indicate that the digital measurements fall within the range of the standard deviation of the respective manual measurements (N = number of measurements per individual fracture, std = standard deviation).

Location	**Fracture ID**	**Manual measurements**					**Digital measurements**		**Difference**	
		Dip Direction [°]	Std	Dip [°]	Std	N	Dip direction [°]	Dip [°]	Dip direction [°]	Dip [°]
Piesberg 1	1	303.0	2.2	70.0	3.3	3.0	302	79	1.0	9.0
	2	308.0	1.2	81.8	0.8	4.0	306	88	2.0	6.3
	3	42.3	1.2	86.3	1.9	3.0	38	84	4.3	2.3
	4	306.5	2.1	78.0	2.0	4.0	303	80	3.5	2.0
	5	32.7	1.9	73.7	4.2	3.0	29	67	3.7	6.7
	6	310.3	1.5	82.8	1.3	4.0	309	86	1.3	3.3
	7	311.3	3.5	85.0	2.8	6.0	307	86	4.3	1.0
	8	310.3	1.3	83.7	2.2	7.0	307	81	3.3	2.7
	9	311.6	1.5	86.6	2.2	5.0	310	88	1.6	1.4
	10	305.0	2.2	78.7	0.5	3.0	299	79	6.0	0.3
	11	302.0	2.4	76.8	4.3	4.0	299	76	3.0	0.8
	12	124.0	1.4	88.0	0.0	3.0	120	87	4.0	1.0
	13	311.9	2.7	85.9	3.0	7.0	311	89	0.9	3.1
	14	313.3	2.5	85.7	2.5	3.0	310	88	3.3	2.3
	Average		*2.0*		*2.2*				*3.0*	*3.0*
	Std		*0.7*		*1.2*				*1.4*	*2.5*

Table A.1 (Continued)

Location	Fracture ID	Manual measurements					Digital measurements		Difference	
		Dip Direction	Std	Dip	Std	N	Dip direction	Dip	Dip direction	Dip
		[°]		[°]			[°]	[°]	[°]	[°]
Piesberg 2	1	339.7	4.2	82.0	0.8	3.0	337	83	2.7	1.0
	2	330.3	3.1	65.0	4.5	3.0	328	68	2.3	3.0
	3	318.8	5.1	89.5	0.5	4.0	316	85	2.8	4.5
	4	62.5	2.3	82.8	4.6	4.0	61	82	1.5	0.8
	5	318.6	3.3	72.4	4.0	7.0	317	71	1.6	1.4
	6	310.0	3.6	81.0	4.3	3.0	312	80	2.0	1.0
	7	311.8	2.7	80.0	6.4	4.0	313	77	1.3	3.0
	8	332.0	7.5	81.6	4.9	5.0	334	82	2.0	0.4
	9	310.3	1.5	75.3	0.4	4.0	311	77	0.8	1.8
	10	310.4	1.5	67.6	3.3	5.0	305	67	5.4	0.6
	11	307.3	3.7	71.7	2.4	3.0	307	68	0.3	3.7
	12	28.0		70.5		2.0	26	72	2.0	1.5
	13	308.0	1.6	83.3	0.8	4.0	310	83	2.0	0.3
	14	71.3	4.0	83.5	4.2	4.0	74	86	2.8	2.5
	15	315.3	5.7	77.6	8.5	5.0	317	75	1.7	2.6
	16	2.0	6.4	75.7	0.5	3.0	2	78	0.0	2.3
	17	319.8	3.8	84.3	5.8	4.0	320	89	0.3	4.8
	18	317.0	1.4	75.7	0.5	3.0	315	79	2.0	3.3
	19	310.5	2.5	78.5	4.9	6.0	310	76	0.5	2.5
	20	307.0	4.4	67.8	4.1	4.0	306	68	1.0	0.3
	21	6.0	5.9	69.7	4.5	3.0	9	68	3.0	1.7
	22	19.3	1.9	70.3	4.1	3.0	20	74	0.7	3.7
	23	311.4	2.2	65.4	0.5	5.0	310	62	1.4	3.4
	24	10.0	1.4	55.3	0.9	3.0	10	53	0.0	2.3
	Average		*3.5*		*3.3*				*1.7*	*2.2*
	Std		*1.7*		*2.2*				*1.2*	*1.3*

Table A.1 (Continued)

Location	Fracture ID	Manual measurements					Digital measurements		Difference	
		Dip Direction [°]	Std	Dip [°]	Std	N	Dip direction [°]	Dip [°]	Dip direction [°]	Dip [°]
Woitzel	1	54.0	1.9	85.1	2.2	7.0	53	83	1.0	2.1
	2	120.3	6.3	58.2	3.6	6.0	122	54	1.7	4.2
	3	108.5		71.0		2.0	104	69	4.5	2.0
	4	157.3	4.2	84.5	4.2	6.0	156	84	1.3	0.5
	5	343.2	3.9	87.2	3.1	5.0	344	88	0.8	0.8
	6	107.0	1.0	87.0	1.7	4.0	109	88	2.0	1.0
	7	178.3	3.5	84.2	2.9	6.0	178	84	0.3	0.2
	8	80.5	1.7	80.0	0.7	4.0	79	79	1.5	1.0
	9	205.0	1.4	57.0	2.1	4.0	203	54	2.0	3.0
	10	116.3	2.5	68.8	6.8	4.0	113	64	3.3	4.8
	11	172.0		70.0		1.0	176	70	4.0	0.0
	12	45.7	1.7	85.7	3.1	3.0	50	86	4.3	0.3
	13	62.3	1.2	87.3	2.4	3.0	62	84	0.3	3.3
	14	130.0	4.0	74.0	2.6	13.0	128	74	2.0	0.0
	15	354.3	4.9	82.8	4.0	4.0	351	84	3.3	1.3
	16	320.6	7.3	88.7	1.7	7.0	322	88	1.4	0.7
	17	121.2	6.5	85.4	4.1	5.0	121	87	0.2	1.6
	18	200.3	2.2	81.5	1.1	4.0	198	81	2.3	0.5
	19	201.4	1.4	83.4	3.4	5.0	200	84	1.4	0.6
	20	216.3	3.0	83.3	3.4	4.0	217	84	0.8	0.8
	21	335.0	7.0	86.0	3.5	6.0	334	81	1.0	5.0
	23	77.3	0.8	86.8	2.4	4.0	77	86	0.3	0.8
	24	114.5	5.9	85.5	0.9	4.0	109	87	5.5	1.5
	25	41.3	4.5	88.2	2.3	6.0	44	88	2.7	0.2
	26	292.0	2.2	89.5	0.8	6.0	291	87	1.0	2.5
	27	323.0	4.6	84.6	4.7	5.0	317	88	6.0	3.4
	28	126.5	3.0	87.8	1.9	4.0	127	88	0.5	0.3
	29	311.5	2.1	86.7	2.2	6.0	308	88	3.5	1.3
	Average		*3.4*		*2.8*				*2.1*	*1.6*
	Std		*1.9*		*1.3*				*1.6*	*1.4*
Overall	**Average**		**3.1**		**2.8**				**2.1**	**2.1**
	Std		**1.8**		**1.8**				**1.5**	**1.8**

A2 Age relationships of faults, veins and fractures

Fractures were exclusively exposed in sub-vertical quarry walls, which inhibited the manual investigation of abutting relationships of the individual fracture sets, and thus the determination of the relative timing. Today, the damage zone of the NNW-SSE striking major normal fault in the east (cf. Fig. 4.3) is not exposed because excavation material from the quarry was deposited in this part of the quarry. The damage zone of a W-E striking normal fault is still exposed and mineralized with quartz veins, representing the damage zone, which occur in a 36 m wide corridor (cf. Wüstefeld et al., 2017a). There is no substantiating evidence that W-E to WNW-ESE striking faults formed before basin inversion in the Upper Cretaceous. This is supported by the partly-sealed W-E striking quartz veins, which are inversion-related based on fluid inclusion analysis (cf. qtz II in Wüstefeld et al., 2017b). An ealier set of quartz veins have a formation age before the Late Jurassic, which may be associated with the NNW-SSE striking fault (cf. qtz I in Wüstefeld et al., 2017b).

The observed relationship between W-E to WSW-ENE striking barren fractures and the large NNW-SSE striking fault in the east implies that the regional stress field was perturbed during or after faulting, which caused the curving of this fracture generation. The barren fractures must have formed post-peak subsidence. That means that the temperature was too low (< 70°C; e.g. Bjørlykke and Egeberg, 1993) and time was too short to drive quartz precipitation kinetics. Based on subsidence models of Senglaub et al. (2005), Wüstefeld et al. (2017a) deduced Paleocene or younger ages for temperature conditions below 60°C in the study area.